STUDY GUIDE

Michael Pretes

University of Alabama

HUMAN GEOGRAPHY
Places and Regions in Global Context FIFTH EDITION

Paul L. Knox   Sallie A. Marston

Prentice Hall
New York   Boston   San Francisco
London   Toronto   Sydney   Tokyo   Singapore   Madrid
Mexico City   Munich   Paris   Cape Town   Hong Kong   Montreal

Editor in Chief, Geosciences and Chemistry: Nicole Folchetti
Acquisitions Editor: Christian Botting
Marketing Manager: Scott Dustan
Project Manager: Tim Flem
Assistant Editor: Jennifer Aranda
Managing Editor, Geosciences and Chemistry: Gina M. Cheselka
Project Manager, Production: Traci Douglas
Operations Specialist: Amanda A. Smith
Supplement Cover Manager: Paul Gourhan
Supplement Cover Designer: Tina Krivoshein
Cover Photo Credit: Sajjad Hussain/AFP/Getty Images

© 2010 Pearson Education, Inc.
Pearson Prentice Hall
Pearson Education, Inc.
Upper Saddle River, NJ 07458

All rights reserved. No part of this book may be reproduced, in any form or by any means, without permission in writing from the publisher.

Pearson Prentice Hall™ is a trademark of Pearson Education, Inc.

The author and publisher of this book have used their best efforts in preparing this book. These efforts include the development, research, and testing of the theories and programs to determine their effectiveness. The author and publisher make no warranty of any kind, expressed or implied, with regard to these programs or the documentation contained in this book. The author and publisher shall not be liable in any event for incidental or consequential damages in connection with, or arising out of, the furnishing, performance, or use of these programs.

> This work is protected by United States copyright laws and is provided solely for teaching courses and assessing student learning. Dissemination or sale of any part of this work (including on the World Wide Web) will destroy the integrity of the work and is not permitted. The work and materials from it should never be made available except by instructors using the accompanying text in their classes. All recipients of this work are expected to abide by these restrictions and to honor the intended pedagogical purposes and the needs of other instructors who rely on these materials.

Printed in the United States of America

10 9 8 7 6 5 4 3 2 1

ISBN-13: 978-0-321-61543-5
ISBN-10:　　　0-321-61543-3

Prentice Hall
is an imprint of

www.pearsonhighered.com

Contents

Preface v

Acknowledgements vii

Chapter 1	Geography Matters	1
Chapter 2	The Changing Global Context	17
Chapter 3	Geographies of Population	34
Chapter 4	Nature and Society	53
Chapter 5	Cultural Geographies	73
Chapter 6	Interpreting Places and Landscapes	90
Chapter 7	The Geography of Economic Development	104
Chapter 8	Agriculture and Food Production	123
Chapter 9	The Politics of Territory and Space	140
Chapter 10	Urbanization	158
Chapter 11	City Spaces: Urban Structure	173
Chapter 12	Future Geographies	185
Appendix 1	Review Questions Answer Key	192
Appendix 2	Writing a Term Paper	207

Preface

This study guide accompanies the fifth edition of the textbook *Places and Regions in Global Context*: *Human Geography* by Paul L. Knox and Sallie A. Marston. For each chapter in the textbook, there is a corresponding study guide chapter comprising four parts:

1. ***Chapter objectives***, which give a general idea of the material covered in the chapter.

2. ***Chapter notes***, which summarize the textbook chapters. The chapter notes cover the major points made in the textbook chapter, and often provide additional examples. The most important key terms are listed in **boldface**. References to the textbook's website, http://www.mygeoscienceplace.com, are found throughout the chapter notes, highlighted by the word website.

3. ***Review questions***. These are fill-in-the-blank, multiple-choice, and true-or-false questions that test factual knowledge of the key concepts and terms in each chapter. Answers to these questions are found in the answer key at the end of the study guide.

4. ***Conceptual questions***. These questions are open-ended essay questions that require you to think about issues raised in the chapter and how they might apply to your own community.

Though the chapter notes in the study guide summarize each chapter in the textbook, they are not a replacement for the textbook. To get the most out of the class, you should read *both* the textbook and study guide chapters, and then use the study guide to refer back to major themes and issues in the textbook. This study guide also contains a short appendix giving some tips on writing term papers as well as some examples of how to list references in term papers.

Acknowledgements

Thanks to Tim Flem and Dan Kaveney at Prentice Hall, and to my students and colleagues, who provided many helpful comments on what they liked and didn't like about study guides.

Michael Pretes teaches in the Department of Geography at the University of North Alabama, in Florence, Alabama.

1

Geography Matters

Chapter Objectives

The objectives of this chapter are to illustrate:

1. Why places matter
2. How geography matters
3. The basic tools required for understanding geography

Chapter Notes

Why Places Matter

Human geography is the study of the spatial organization of human activity and of people's relationships with their environments, and with the places and spaces that they are a part of. Places are unique, yet they are interconnected. Places are also dynamic, with changing features and boundaries, which result from both environmental and human factors. Experience and knowledge of other places provides alternative ideas for our own culture in either a positive or negative way. Different values, attitudes, and behaviors are shaped by different places. Places can also be socially constructed, in that their meaning is shaped by society. Places also influence people's physical well-being, opportunities, and lifestyles. Places can also be symbolic—think about how Mount Rushmore or the Washington Monument shape and reflect Americans' ideas about democracy, for example. Places can also be the sites of innovation and change. They can accept, alter, or resist new ideas and technologies.

Places are *interdependent*; that is, they are linked to each other and depend upon each other. Consider the foods you eat for breakfast: you might have grapefruit from Florida, bananas from Guatemala or Ecuador, cereal from the American Midwest, milk from your local dairy, and coffee from Brazil or Indonesia. You depend on these places for your food, and they depend on you as a customer. This interdependence is true on both a local and global level. Things that happen throughout the world can influence your daily life (like the price of coffee or gasoline). Things that happen locally can influence the rest of the world (for example, a film or book set in your area, or the crops, manufactures, or services that your region provides). Places are dynamic: they are always changing, adjusting themselves to the world situation.

Different aspects of human geography can be understood and analyzed at different spatial scales. Scale is a tangible partitioning of space. Different scales represent different confluences of geographic processes. **World regions** are large-scale geographic divisions based on continental and physiogeographic settings that contain major groupings of people with broadly similar cultural attributes—Europe or Latin America, for example. Within these regions are national **states**. States are independent political units with territorial boundaries recognized by other states (*nations* are groups of people sharing similar cultural characteristics, such as Finns or Samoans). **Supranational organizations** are collections of individual states with a common goal, such as the European Union. Smaller regions are also found within world regions and states. Geographic scales can get progressively smaller, working down from the level of world region to the scale of the home or even the human body.

Places are not just outcomes of geographical processes; they are part of these processes themselves. Places are created by people responding to opportunities and constraints presented by their environments. It is a *two-way process*, as people both create and modify places while at the same time being influenced by them.

Interdependence in a Globalizing World

Globalization is another way of expressing the interconnectedness between all parts of the world. We can now speak about *one* world economy, in which all countries take part. Their place within that single economy has much to do with their level of development, as we will see later in other chapters. Globalization is increasing, largely due to improvements in transportation and communication technology. The fact that we can travel between continents in a few hours in jet airplanes, or communicate across continents by e-mail in a few seconds has totally changed our relationship with the rest of the world. These improvements in technology also mean that human beings are in a position, as never before, to completely alter the global environment. Pollution, global warming, and deforestation are just a few of the ways in which human beings have negatively affected the natural environment that surrounds them. The fact that we live in a globalized world doesn't necessarily mean that all places in the world will one day be all alike. People have resisted globalizing trends, and have sought refuge in the past. Think about "Main Street" as portrayed in Disneyland or Walt Disney World, as well as many other examples of nostalgia for small town life. Nationalist movements, such as those in Québec, Serbia, or Iran, are also examples of resistance to a global culture.

Studying Human Geography

Studying geography helps us to understand the world around us, and why people and cultures are the way they are. Geography is generally divided into two main branches: **physical**, which concerns climate, landforms, and vegetation; and **human**, which includes human society and its relationship to the environment. Geography is unique in that it is both a natural *and* social science. Human geography can be studied from two different perspectives: regional and systematic. **Regional** geography focuses on regions, which are analytical constructs based on common features such as landscape, language, religion, and political and economic systems. Latin America, the Middle East, and East Asia are examples of regions, as are Scandinavia, New England, and the American West.

Regional geographers study everything about a region—climate, landforms, language, culture, etc.—in order to see what makes it distinct. Systematic geography approaches the study of geography by topic.

Geographers use many tools and techniques. **Remote sensing** is an advanced technique for cartography (map-making), using images of Earth collected by satellites or by other forms of aerial photography. Google Earth is an example of a site that provides satellite images of Earth. **Geographic information systems (GIS)** are advanced computer maps that superimpose multiple layers, showing more kinds of data and the interactions between them than is possible with traditional map forms. They are often used for **geodemographic research** to create profiles of regions to assist marketing companies, for example.

Spatial analysis is the study of the space around us, and is usually divided into five concepts:

Location: we can refer to a place by its name (San Francisco, Lake Superior), which is called *nominal location*, or by its exact location (by lines of **latitude**, which run east–west around the Earth, and lines of **longitude**, which run north–south up and down the Earth, which is called *absolute location*. Latitude and longitude are now very easy to determine with the new technology of the **Global Positioning System (GPS)**, which consists of 21 satellites that can broadcast locational information to anyone with a GPS receiver. Location can also be relative, which refers to a place's **site** (its physical attributes: soil, vegetation, water, etc.) and its **situation** (the location of the place relative to other places). San Francisco, California, is located on the tip of a peninsula surrounded by water on three sides—that's its site. San Francisco is also located by the Pacific Ocean in Northern California—that's its situation. A final form of location is **cognitive**, also known as mental maps. This refers to how we perceive a place. To consider Hawai'i as a tropical paradise is a form of cognitive location.

Distance: the geographer Waldo Tobler noted that "everything is related to everything else, but near things are more related than distant." What does this mean? But first, how do we measure distance? We can measure distance in *absolute* terms such as miles or kilometers, or in *relative* terms, such as the number of hours or the cost of travel.

Distance can also be **cognitive**; that is, it can also be a mental perception. People in Europe might think that driving 60 miles to work is an outrageous distance, but for people in Los Angeles 60 miles might seem short. People tend to avoid going long distances for things (unless they judge the benefit to be worth the extra travel): this is known as the **friction of distance**. **Distance-delay function** refers to the rate at which a particular activity diminishes the further one moves away from a central location. People generally try to find the optimal relationship between distance and the benefit they receive from it, such as traveling to a distant store if prices are much lower, but staying nearer home if prices are equal or not low enough to travel that far.

Space: Space can also be measured in absolute, relative, and cognitive terms, just like the measurement of distance. **Topological** space refers to the connections between particular points.

Accessibility: refers to the opportunity for contact or interaction from a given location in relation to other locations. We commonly express this concept by saying that some places are more accessible than others. Distance is only one of the factors of accessibility. A given place may be closer to you than another, but not as accessible (because of a bad road or no road, for example). Your geography instructor may or may not be accessible, depending on how many office hours he or she keeps, and whether they are at convenient times for you.

Spatial interaction refers to the flows and movements between places. Spatial interaction itself has four basic concepts within it. In order for spatial interaction to take place between two locations, there must be some reason for it; this is called *complementarity*. Often this is based on factors of supply and demand: the two places have something to exchange or trade with each other. If the United States produces computers and Ecuador produces bananas, then we have the basis for trade in these two items. The United States is not likely to develop a banana industry because of climate and because the tropical lands of the United States are best used for other purposes. Neither is Ecuador likely to develop a major computer industry, largely because it cannot compete against countries like the United States, where an economy of scale exists. An **economy of scale** refers to cost advantages that result from high-volume production. If it costs so many dollars to produce 100,000 computers, it does not cost much more to produce 150,000 because the infrastructure for production is in place. *Transferability* refers to the ability to move an item from one place to another. This can reflect costs and the ability of the item to make the move (it might be too perishable to be transported long distances). Transferability is constantly changing, especially given the introduction of new technologies. The idea of a shrinking world, in which transportation and communication times are much faster, is called **time–space convergence**. Europe is "closer" to the United States now that we have jet aircraft than it was in the days of steamships. **Intervening opportunity** refers to alternatives that may affect the transferability between two places. If Florida started producing bananas, the United States might stop importing them from Central and South America. Florida bananas would be an intervening opportunity. **Spatial diffusion**: is the way things spread out geographically. Many things—musical styles, diseases, inventions—tend to originate in one place and then spread to others. In *expansion diffusion*, things spread across space and time by carriers that remain fixed. In *relocation diffusion*, things spread by carriers that migrate and take the new thing with them. In *hierarchical diffusion*, things spread downward in a given hierarchy.

Landscape is the product of human interaction with the environment. **Ordinary landscapes** (or **vernacular landscapes**) are the everyday landscapes that people create in the course of their lives. Parking lots, suburban housing developments, and farms are all examples of ordinary landscapes. **Symbolic landscapes** are representations of particular values and/or aspirations that the builders and financiers of those landscapes want to impart to a larger public. For example, many government buildings in the United States, such as court houses, are built in a neoclassical style, which symbolizes not only power and authority, but also the Greek and Roman classical ideals of democracy and republicanism. People can establish bonds with each other by sharing a sense of place. A **sense of place** is the feeling evoked among people as a result of the experiences and memories that they associate with a place, and to the symbolism that they attach to it. We can identify people as "locals" or as "outsiders." Common bonds may exist between people who consider themselves "Southerners" or "Texans," bonds that are not formed with people who do not share this same sense of place. The sense of place may be reflected in shared dress codes, speech patterns, public behavior, and so on. Texans, for example, may recognize each other by their accent, indicating who is a Texan and who is not; that is, who shares the sense of Texas space and who does not.

Developing a **geographical imagination** is the goal of geographic study. A geographical imagination is the capacity to understand the changing patterns, changing processes, and changing relationships among people, places, and regions. If you look around your own neighborhood with a geographical imagination you will understand why things are the way they are.

Geography is an applied discipline and can make a direct contribution to society. Examples of these contributions include international affairs, locating public facilities, marketing and the location of industry, legal issues, disease ecology, urban and regional planning, and economic development.

Review Questions

Some of the answers to these questions may be found only in the textbook, and not in the study guide.

1. The rate at which places move closer together in travel or communication time or costs is:

   a) spatial diffusion

   b) distance-decay function

   c) friction of distance

   d) time–space convergence

2. _____ are psychological representations of locations that are made up from people's individual ideas and impressions of these locations.

8 Chapter 1

3. _____ is the opportunity for contact or interaction from a given point or location, in relation to other locations.

4. The rate at which a particular activity or process diminishes with increasing distance is:

    a) distance-decay function

    b) friction of distance

    c) spatial diffusion

    d) time–space convergence

5. True or False: Economies of scale are cost advantages to manufacturers that accrue from high-volume production, since the average cost of production falls with increasing output.

6. Groups of areal units that have a high degree of homogeneity in terms of particular distinguishing features are _____.

7. _____ uses census data and commercial data (such as sales data and property records) about the populations of small districts in creating profiles of those populations for market research.

8. _____ is a system of satellites which orbit the Earth on precisely predictable paths, broadcasting highly accurate time and locational information.

9. Patterns of interaction among family members, at work, in social life, in leisure activities, and in political activity are known as _____.

10. The branch of geography that deals with the spatial organization of human activities and with people's relationships with their environment is:

    a) physical geography

    b) human geography

c) regional geography

d) national geography

11. Fixed social capital constitutes the _____ of society and includes things like roads, highways, and schools.

12. A _____ is a larger-sized territory that encompasses many places, all or most of which share similar attributes in comparison with the attributes of places elsewhere.

13. The increasing interconnectedness of different parts of the world through common processes of economic, environmental, political, and cultural change is known as _____.

14. _____ is the geographer's equivalent of scientific classification, with individual places or areal units being the objects of classification.

15. The assertion by a government of a country that a minority living outside its formal borders belongs to it historically and culturally is _____.

16. True or False: Latitude is the angular distance of a point on the Earth's surface, measured north or south from the equator, which is 0°.

17. The concept of _____ refers to the feelings evoked among people as a result of the experiences and memories they associate with a place.

18. The distance that people perceive to exist in a given situation is known as _____.

19. The study of many geographic phenomena in terms of their arrangement as points, lines, areas, or surfaces on a map is _____.

20. What are the five concepts that geographers use in analyzing social patterns and distributions?

_____

_____

21. Landscapes that represent particular values or aspirations that the builders and financiers of those landscapes want to impart to a larger public are known as:

    a) ordinary landscapes

    b) boring landscapes

    c) symbolic landscapes

    d) vernacular landscapes

22. People's _____ is the sense that they make of themselves through their subjective feelings based on their everyday experiences and wider social relations.

23. True or False: Supranational organizations are collections of individual states with a common goal.

24. The everyday landscapes that people create in the course of their lives together are _____.

25. _____ is the study of the ways in which unique combinations of environmental and human factors produce territories with distinctive landscapes and cultural attributes.

26. _____ is geographers' shorthand for movement and flows involving human activity.

27. _____ is a term used to describe situations in which different religious or ethnic groups with distinctive identities coexist within the same state

boundaries, often concentrated within a particular region and sharing strong feelings of collective identity.

28. Large-scale geographic divisions based on continental and physiographic settings that contain major groupings of peoples with broadly similar cultural attributes are:

    a) places

    b) nations

    c) states

    d) world regions

29. True or False: Distance-decay function is the way that things spread through space over time.

30. The deterrent of inhibiting effect of distance on human activity is known as _____.

31. The collection of information about parts of Earth's surface by means of aerial photography or satellite imagery designed to record data on visible, infrared, and microwave sensor systems is known as _____.

32. Extreme devotion to regional interests and customs is known as:

    a) regionalism

    b) regionalization

    c) nationalism

    d) sectionalism

33. The location of a place relative to other places and human activities is its _____.

34. _____ is the capacity to understand changing patterns, changing processes, and changing relationships among people, places, and regions.

35. Space defined and measured in terms of people's values, feelings, beliefs, and perceptions about locations, districts, and regions is _____.

36. _____ are independent political units with territorial boundaries that are recognized by others.

37. Space that is defined and measured in terms of the nature and degree of connectivity between is known as:
    a) spatial diffusion
    b) topological space
    c) time–space convergence
    d) regionalization

38. Integrated computer tools for the handling, processing, and analyzing of geographical data are known as _____.

39. The usefulness of a specific place or location to a particular person or group is its _____.

40. _____ occur when external economies and local economic linkages are limited to firms involved in one particular industry.

41. The concept of _____ refers to the feelings evoked among people as a result of the experiences and memories they associate with a place.

42. _____ refers to the angular distance of a point on the Earth's surface north or south of the equator, while _____ refers to points east or west of a line passing through Greenwich, England.

43. The process by which such things as the sport of soccer or the HIV virus spread through space over time is known as _____.

Conceptual Questions

1. What is meant by "globalization"? What evidence do you see of this process in your own area?

_____
_____
_____
_____
_____
_____
_____
_____
_____
_____

2. What symbolic landscapes are present in your community or region? What makes them symbolic? What do they symbolize? What message is being sent through these landscapes and who is sending it?

_____
_____
_____
_____
_____
_____
_____
_____

3. Choose a city or place in your area and describe both its *site* and *situation*. What is the relationship and connection between these?

4. Discuss some of the ways that geographic information systems (GIS) can contribute to geographic analysis.

Geography Matters 15

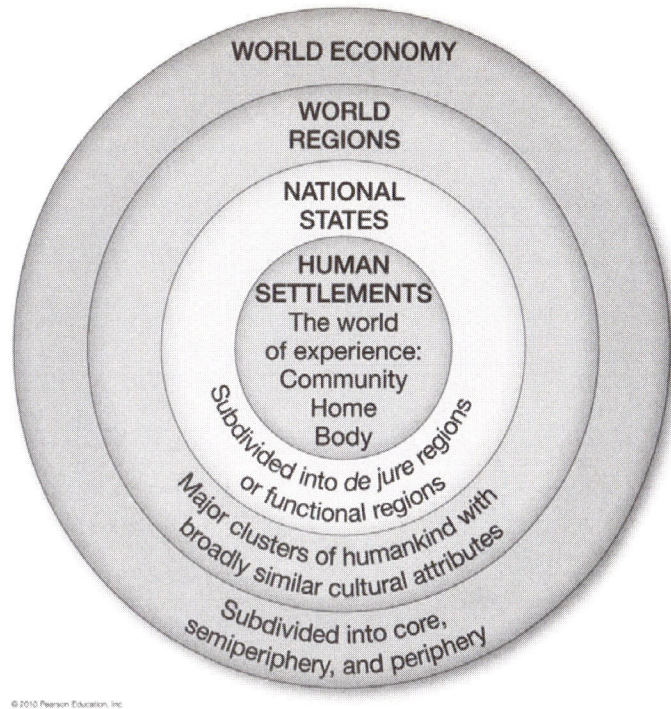

Figure 1.4

5. Look at Figure 1.4, which illustrates spatial scales. Can you think of examples for each of the scales shown in the concentric circles?

_____
_____
_____
_____
_____
_____
_____
_____
_____
_____

© 2010 Pearson Education, Inc.

6. Draw a map of your area. To what extent is this map a cognitive image?

2

The Changing Global Context

Chapter Objectives

The objectives of this chapter are to illustrate:

1. Geographic expansion, integration, and change
2. Industrialization and geographic change
3. Forces that organize the periphery
4. The fast world and the slow world

Chapter Notes

The Premodern World

The study of human geography requires an understanding that different places and regions are part of a constantly changing global system. Each place reflects its own history. **Minisystems** are societies with a single cultural base and a reciprocal economy, in which individuals share their excess products with each other. Many, but not all, Native American societies before European contact could be considered minisystems. The agricultural revolution, or the beginning of agriculture, began around 9000 to 7000 BC. An early agricultural technique, which is still practiced in parts of the world, is "swidden" or **slash-and-burn** agriculture. In this method, plants are cropped close to the ground, left to dry, then set on fire. The resulting ash provides nutrients for the soil.

Practices, products, and ideas generally originate from a specific location on the globe. This specific source area is known as a **hearth area**. Agriculture, for example, is a practice that did not appear everywhere simultaneously. Agriculture instead originated in a few distinct places, and then spread or diffused to other regions. In addition, certain agriculture products originated in certain hearth areas and gradually diffused throughout the world. A good example would be coffee, which originated in Ethiopia and then spread to Arabia, then to Europe, and then to European colonies in South America. The agricultural revolution probably took place in four distinct hearth areas at approximately the same time in history. These four hearth areas were: the Middle East, especially in Mesopotamia or the "Fertile Crescent"; South Asia, especially in the floodplains of major rivers; China, along the Yuan River; North and South America, especially in Mexico and the American Southwest, and on the western slopes of the Andes Mountains in South America.

The idea of a **world-system**, suggested by Immanuel Wallerstein, views the world as consisting of one single economic system, in which each country or region takes part in different ways. The world-system is an *interdependent* system, in which each country and region is linked to the others by economic and political competition. The agricultural revolution allowed for increased food production, increased population, changes in social organization, and craft production and trade. These were the preconditions for the emergence of **world-empires**, which are minisystems that have been absorbed into a common political system while retaining their fundamental cultural differences. The Roman Empire is one example of a world-empire. The emergence of world-empires allowed for the development of *urbanization* and **colonization**. Colonization was an indirect consequence of the **law of diminishing returns**, which notes that productivity tends to decline after a certain point, even with the addition of capital and labor to a given resource base. For example, adding fertilizer as well as labor to a piece of land will increase its productivity; but eventually a point is reached when the additional fertilizer and labor has no effect, or even a negative effect.

Before the 1400s, the world consisted of a number of world-empires, one of which is shown in Figure 2.4 in the textbook. Some of these were *hydraulic societies*, which were based around rivers or other waterways. Some scholars have argued that hydraulic societies were the first true states, as they required a permanent political elite to direct the massive irrigation projects required to feed the population. In general, the most important world-empires had some water connection, which allowed for easier trade with other world-empires. The growing cities of each world-empire were the centers of activity and trade. Each city controlled a **hinterland**, which is the sphere of economic influence of a town or city.

Mapping a New World Geography

The world-system began in the 1400s, when Europeans started exploring and settling beyond their home region. European expansion brought about the exchange of ideas, technologies, and resources between regions that previously had little to do with each other. Those regions of the world that had not been absorbed into the world system were known as the **external arena**. Within the past few hundred years the world-system grew even stronger, now encompassing the entire world. Geography and **cartography** had a major influence on European territorial expansion. In the late 15th century, Europeans sought new trade routes as well as scientific knowledge, and in doing so used, improved, and compiled maps, thus increasing their knowledge of the world. The textbook's website has a link to the ancient map collection at the University of Texas, which depicts a number of historical maps. Notable 15th-century European explorers included Bartholomeu Dias (Africa), Christopher Columbus (Caribbean), Vasco de Gama (India), and Pedro Cabral (Brazil). Within a few years Europeans had reached much of the world, and in the early 16th century, Magellan made the first circumnavigation of the globe.

Geographical knowledge acquired during these voyages of discovery was crucial to the expansion of European political and economic power. During the European Renaissance cartographers developed many new ways of depicting the Earth as a flat surface through new **map projections**. Geographic knowledge and new map projections assisted the European project of colonization. European expansion abroad and the exploitation of natural resources outside of Europe allowed Europe to develop as a core region. Europeans brought back valuable resources such as gold and silver from other parts of the world, and established **plantations**—large landholdings specializing in the production of a single crop for sale as opposed to local consumption—in their colonies. Europe was also able to practice **import substitution**, in which products previously imported into Europe could now be made locally, making Europe less dependent on imports.

The world-system is organized hierarchically, with some regions at the "top" and others at the "bottom"—take a look at the Core-Periphery Model link on the textbook website for a schematic depiction of this organization. There are three kinds of region: **Core regions** are those that dominate trade, control the most advanced technologies, and have high levels of productivity within diversified economies. The core states of today include most of western and central Europe, the United States and Canada, Japan, and a few other places such as Australia. Some of these core states achieved their core position through **colonialism**, which is the establishment and maintenance of political and legal domination by a state over a separate and alien society. Colonialism helped the core states exploit resources in other parts of the world, leading to the political and economic growth of the core states. **Peripheral regions** are those with undeveloped or narrowly specialized economies with low levels of productivity. Peripheral states are sometimes known collectively as the "Third World" or "Least Developed Countries." Their economies and governments are often influenced and even dominated by the core and semiperipheral states. Examples of peripheral states include Ethiopia, Nepal, Bolivia, and Guatemala, among many others. **Semiperipheral regions** are states and regions that are in between core and peripheral states in terms of their economic and political development and freedom from domination. The economies of the semiperipheral states are influenced by the core states, but at the same time the semiperipheral states can also influence and dominate the peripheral states below them in the hierarchy. Examples of semiperipheral states today would include Mexico, Brazil, India, and Taiwan.

We can trace the establishment of geography as a formal academic discipline to such 18th- and 19th-century writers as Immanuel Kant, Alexander von Humboldt, Karl Ritter, and Friedrich Ratzel. The American geographer Carl Sauer added further sophistication to the field in the 1920s by arguing that landscape should provide the focus for the scientific study of geography. European geography before the 20th century often (but not always) served nationalist ends. European geographers tended to view the world as though Europe was at the center. Much of European geography in the 19th century was influenced by several concepts: **ethnocentrism**, the belief that one's own culture is superior to all others; **masculinism**, the assumption that the world is largely shaped by and for men (and not women); and **environmental determinism**: the belief that human activity is completely shaped by the environment. The belief that Europe emerged as a world power because of its mild climate is an example of environmental determinism. Geography in the 20th century and today largely rejects all of the above concepts.

The Industrial Revolution, which began in the late 1700s, introduced many new technologies and modes of production, allowing for further concentration of economic power in core regions. Industrialization permitted the low-cost manufacture of many standardized products. Europe, the United States, and Japan all participated in industrialization within the past several hundred years, allowing their economies to out-compete production elsewhere. In addition to industrialization, the development of transportation technology aided in the development of the core regions. Canals, steamboats, and railroads all facilitated increased transportation and communication within the core countries, allowing them to colonize their own hinterlands in a similar manner to the way in which they colonized external territories. Canals, steamboats, and railroads all made it easier and cheaper to transport various kinds of products. Later developments in transportation technology and in related fields included tractors, trucks, road building, and spatial reorganization. These developments had parallel results to that of canals, steamboats, and railroads. In addition, the development of trucks allowed for even greater penetration of the hinterlands, to places not served by water or rail connections.

At the same time as core regions were developing and exploiting their own hinterlands, they were also developing and exploiting territories beyond their borders through the processes of colonialism and imperialism. As the economies of the core powers grew, they needed new supplies of raw materials and foodstuffs as well as new places in which to sell their manufactured goods. One of the outcomes of this process was the **division of labor**, which meant the specialization of different people, regions, and countries in certain kinds of economic activities. The logic of the division of labor required that colonies produce products that were needed in the core, that they held a comparative advantage in these products, and that the products did not duplicate or compete with things produced in the core. **Comparative advantage** means that places or regions will specialize in activities for which they have the greatest advantage in productivity relative to other regions. For example, cotton can be grown in many regions, but India held a comparative advantage in cotton production, because India could produce it more efficiently and cheaply—due to climate and cheap labor—compared to many other places. The result of these colonial practices was a very narrow specialization in most colonies. For example, Ghana depends on cocoa, and very little else, while Kenya depends on tea and coffee, and very little else. These places are locked into a peripheral position, because they depend on very few export products that they *must* sell abroad, whether demand and prices are favorable or unfavorable. The single most important innovation behind the international division of labor was the development of the metal-hulled, ocean-going steamship. This allowed products to be transported swiftly and cheaply across long distances, from colonies to the core centers of Europe and the United States.

Many countries took part in the processes of colonialism and imperialism. Some parts of the world, such as Africa, were almost completely colonized and controlled by European powers—the textbook's website has an Imperialism and Colonialism link that examines the European colonization of Africa. The periphery became almost totally dependent on European and North American manufactured products, as well as capital, shipping, managerial expertise, financial services, and news and communications. Consequently, core cultural ideas also penetrated the colonized and dependent areas, in such forms as language, education, science, religion, architecture, and planning. After World War II, the world-system began to change. The imperial world order began to disintegrate as many colonies started achieving their independence. The Soviet Union and China also emerged as major powers, both pursuing an alternative path to development—communism. Many scholars began to identify three new "worlds" within the world system. The "First World" was the industrialized countries or core, the "Second World" was the communist countries including the Soviet Union and China and their allies, and the "Third World" was the less-developed countries of Africa, Asia, and Latin America.

But the emergence of three "worlds" did not mean that that world-system was breaking down. On the contrary, it indicated that the world-system was more interdependent than before. Direct colonialism, in which core countries governed their colonies, was replaced by **neocolonialism**, in which core countries took an indirect, but more complex, involvement in the economic and political affairs of the former colonies through financial assistance and trade. Another sign of growing interdependence in the world-system was the increasing role of **transnational corporations**, which are companies with investments and business operations that span international boundaries and with facilities in many countries. Many of these corporations, such as Exxon Mobil, General Electric, Nestlé, and others, have substantial investments in developing countries and strongly influence the economies of those nations. Go to the textbook's website to see the Age of Globalization link, which examines the power of transnational corporations today.

Contemporary Globalization

The world-system has become increasingly complex, as international links and connections have expanded. These networks are often **commodity chains**, which are networks of labor and production processes beginning with the extraction of raw materials and ending with the delivery of a finished product. Transnational corporations often control many aspects of the commodity chain.

The globalization of the past twenty-five years has been caused largely by four basic factors: 1) A new international division of labor. This new international division of labor has three facets: the decline of the United States relative to Europe and Japan; manufacturing has moved from the core to the semiperiphery and even to the periphery; and the emergence of new specializations, such as high-tech industries and information services, in the core regions. 2) Internationalization of finance. There is now essentially one global banking system and one globally-integrated financial market. 3) A new technology system based on innovations such as robotics, microelectronics, computerization, and biotechnology. Satellites, fax machines, and the internet have all changed and speeded up communication between places. 4) The growth of consumer markets. There has been an increased demand for high-status international products and consumer goods, many of which are luxury items.

With the changes mentioned above, it no longer seems correct to speak of a First World, Second World, and a Third World. Instead, we might now speak of a Fast World and a Slow World. The **Fast World** refers especially to the core regions, to people, regions, and places directly involved, as producers and consumers, in transnational industry, modern telecommunications, materialistic consumption, and international news and entertainment. The **Slow World** refers to people, regions, and places where these things are limited. The fairness of this distribution of burdens and benefits is reflected in the concept of **spatial justice**. The end result of all of the above is that places and regions still matter: they have not been obliterated by globalization. In fact, globalizing processes may actually support local identities, an argument made on the National Geographic Globalization link found on the textbook's website. Each place and region reflects its history and the legacy of changes in the world-system.

Review Questions

Some of the answers to these questions may be found only in the textbook, and not in the study guide.

1. The establishment and maintenance of political and legal domination by a state over a separate and alien society is known as _____.

2. The three divisions of the world-system are:

_____

3. True or False: Imperialism is the physical settlement in a new territory of people from a colonizing state.

4. A network of labor and production processes beginning with the extraction or production of raw materials and ending with the delivery of a finished commodity is a:
    a) commodity chain
    b) minisystem
    c) comparative advantage
    d) hinterland

5. The assumption that the world is, and should be, shaped mainly by men for men is known as _____.

6. Regions that dominate trade, control the most advanced technologies, and have high levels of productivity within diversified economies are known as:
    a) peripheral regions
    b) semiperipheral regions
    c) core regions
    d) hinterlands

7. _____ is the specialization of different people, regions, or countries in particular kinds of economic activities.

8. True or False: A plantation is a large landholding that usually specializes in the production of one particular crop for market.

9. The doctrine holding that human activities are controlled by the environment is _____.

10. True or False: Masculinism is the attitude that one's own race and culture are superior to those of others.

11. Regions of the world not yet absorbed into the modern world-system are called the _____.

12. From which four hearth regions did agricultural practices spread?

_____

_____

13. _____ refers to people, places, and regions directly involved, as producers and consumers, in transnational industry, modern telecommunications, materialistic consumption, and international news and entertainment.

14. Geographic settings where new practices have developed, and from which they have subsequently spread are known as:

   a) hinterlands

   b) core regions

   c) hearth areas

   d) peripheral regions

15. The sphere of economic influence of a town or city is its _____.

16. True or False: The deliberate exercise of military power and economic influence by powerful states in order to advance and secure their national interests is known as colonization.

17. The first wave of industrialization in Europe occurred in the country of _____.

18. _____ is the process by which domestic producers provide goods or services that formerly were bought from foreign producers.

19. The tendency for productivity to decline, after a certain point, with the continued application of capital and/or labor to a given resource base is:

    a) law of diminishing returns

    b) neocolonialism

    c) environmental determinism

    d) comparative advantage

20. A society with a single cultural base and a reciprocal social economy is a:

    a) core region

    b) minisystem

    c) fast world

    d) external area

21. In general terms, the core of the world economy is now structured around the three centers of: _____

22. Economic and political strategies by which powerful states in core economies indirectly maintain or extend their influence over other areas or people is _____.

23. The first phase of internal economic expansion in core countries was based on what type of transportation system/technology? _____.

24. Regions with undeveloped or narrowly specialized economies with low levels of productivity are:

    a) core regions

    b) semiperipheral regions

    c) peripheral regions

    d) technology regions

25. A company with investments and activities that span international boundaries and with subsidiary companies, factories, offices, or facilities in several countries is a _____.

26. _____ are services that enhance the productivity or efficiency of other firms' activities or that enable them to maintain specialized roles.

27. Regions that are able to exploit peripheral regions but are themselves exploited and dominated by core regions are called _____.

28. _____ is a system of cultivation in which plants are cropped close to the ground, left to dry for a period, and then ignited to release nutrients into the ground.

29. True or False: The fast world refers to people, places, and regions whose participation in transnational industry, modern telecommunications, materialistic consumption, and international news and entertainment is limited.

30. The fairness of the distribution of society's burdens and benefits, taking into account spatial variations in people's needs and in their contribution to the production of wealth and social well-being is _____.

31. Clusters of interrelated energy, transportation, and production technogies that dominate economic activity for several decades at a time are:
    a) minisystems
    b) world-empires
    c) technology systems
    d) hinterlands

32. True or False: Comparative advantage is the principle whereby places and regions specializing in activities for which they have the greatest advantage in productivity relative to other regions or for which they have the least disadvantage.

33. The system of practical and theoretical knowledge about making distinctive visual representations of Earth's surface in the form of maps is known as:
    a) imperialism
    b) spatial justice
    c) cartography
    d) map projections

34. True or False: World-empire refers to minisystems that have been absorbed into a common political system while retaining their fundamental cultural differences.

35. An interdependent system of countries linked by economic and political competition is a _____.

36. By the end of the 19th century, the core of the world system had expanded to include the countries of _____ and _____.

37. Using Figure 2.19 (and other maps if necessary), can you name five core countries, five semiperipheral countries, and five peripheral countries?

_____
_____
_____

30   Chapter 2

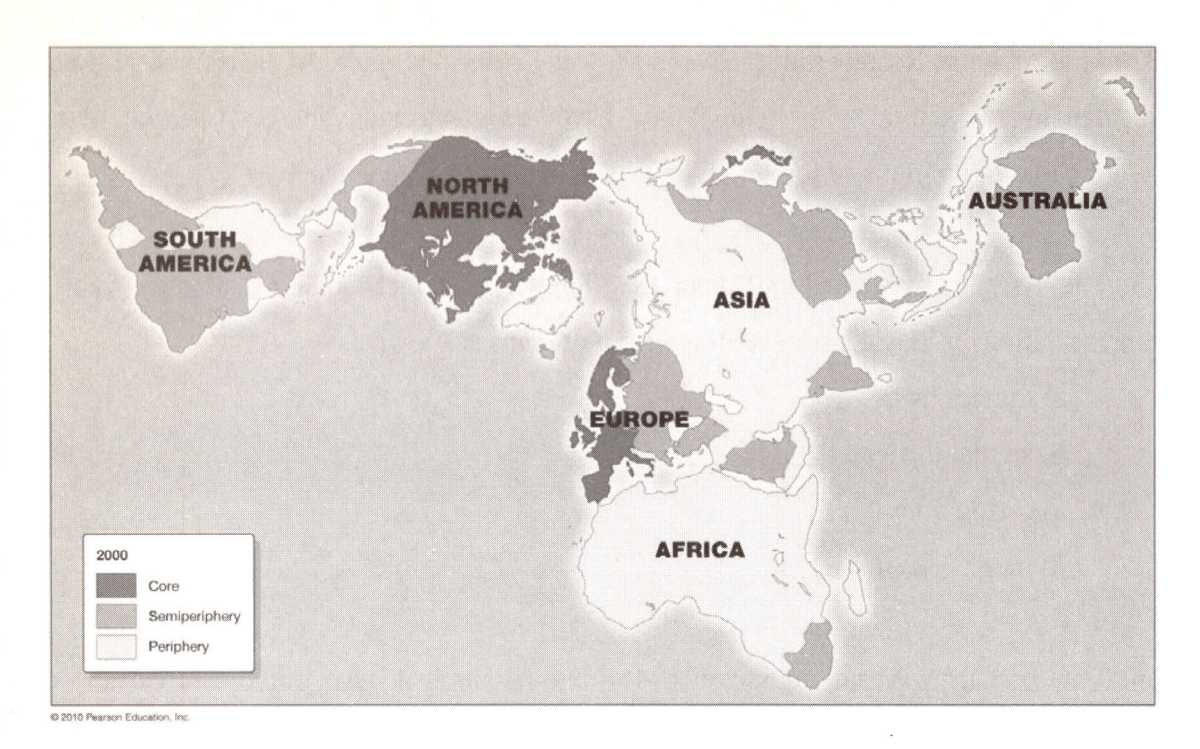

Figure 2.19

Conceptual Questions

1. What were some of the major factors that accounted for the rise of a world-system in the 1500s? Why was Europe so critical in the development of the world-system?

2. Describe the three basic components of the world-system, giving examples of states in each category.

3. Why is the Netherlands a richer and more economically developed country than Guatemala? How does the world-system model help to explain regional differences in levels of economic development?

4. Are there any minisystems left in the world today? If you think so, give some examples and explain why you think these minisystems have not merged into the world-system.

32 *Chapter 2*

5. What effects did innovations in transportation technology—canals, steamboats, railroads, and trucks and automobiles—have on the growth of core regions?

6. What factors are largely responsible for the increase in globalization in the past 25 years? Can you give examples to illustrate these factors?

7. What is the difference between the Fast World and the Slow World? Why does it no longer seem appropriate to speak of First, Second, and Third Worlds?

3

Geographies of Population

Chapter Objectives

The objectives of this chapter are to:

1. Examine the national census and understand its limitations
2. Investigate population distribution and structure
3. Explore various population dynamics and processes
4. Understand population movement and migration
5. Discuss current population policies and debates

Chapter Notes

The Demographer's Toolbox

Demography, the study of the characteristics of human populations, involves many academic disciplines. Geographers who study population tend to emphasize the spatial patterns of human population, including the implications of these patterns and reasons for them. Geographers also look at how places shape populations and how populations shape places. Information about populations can be gathered in a variety of ways, and two of the most important are census data and vital records. A **census** is a count of the number of people in a country, region, or city, though most censuses also gather other personal information, such as marital status and income, in addition to simply counting people. Thus, a census can provide a great deal of information about a population. The United

States conducts a census every ten years, in years ending with a zero. You can get extensive data about the United States census from its website—see the link in the textbook's website, or try www.census.gov. **Vital records** are another source of population data. Vital records include things like birth records, death records, marriage and divorce records, and information about the incidence of certain diseases. This information is collected and recorded by different local, county, and state agencies and sometimes by other organizations such as schools and hospitals.

Census data is subject to a number of potential problems. Censuses are expensive and complicated to administer, especially in very large countries like the United States, and therefore can only be conducted every few years. A census conducted in 2000, for example, is not very up-to-date in 2009. Moreover, different countries conduct their censuses in different years, making cross-national comparison difficult. An additional problem with census data is the use of ethnic or linguistic categories, much of which relies on self-identification. Sometimes broad categories such as "Hispanic" will not truly reflect the diversity of the categories that they attempt to measure.

Population Distribution and Structure

Population is not distributed evenly around the world. Some parts of the world are heavily and densely populated, while other parts are almost empty of people. Environmental/physical and historical/social factors account for these differences. Environmental and physical factors include such things as degree of accessibility, topography, soil fertility, climate and weather, water quality, and availability of various natural resources. Because of different natural resource endowments, some parts of the Earth are able to support larger populations. Historical and social factors also partly account for the uneven distribution of the world's population. Some areas are heavily populated because they were centers of trade or colonial administration. Both the Population Reference Bureau and the Population Research Institute have extensive data on the world's population—see the textbook's website for the link. Figure 3.2 in the textbook illustrates the distribution of the world's population. In looking at this figure, you will notice that population is heavily concentrated in Asia and Europe. China and

India together have about 21 of the world's total population. Almost all of the world's inhabitants live on 10 percent of the land, on land that is generally characterized by a temperate climate, is relatively low-lying or flat, and contains fertile soils.

Population distribution is often understood in terms of density, which is the ratio of people to some other unit. **Crude density** (also known as arithmetic density) is perhaps the most common measure of population density. Crude density is simply the number of people in an area divided by the total land area. Crude density is a fairly simple measure of population distribution, but it does not indicate anything about the type of land being measured and how useful it is for supporting a population. Other measures, such as nutritional density and agricultural density, help to do this. **Nutritional density** is the ratio between the total population and the amount of land under cultivation in a given area, while **agricultural density** is the ratio between the number of agriculturalists or farmers per unit of farmable land in a given area. Both of these measures reflect the ability of the land to support a population's food needs.

Another important aspect of population geography is the composition of the population. Population geographers are interested in knowing the breakdown of a population by the numbers of males and females, young and old people, and economically active people, for example. Some countries have a high proportion of older people, while others have a high proportion of children. Many core countries have a large **baby boom** population, which refers to people born between the years 1946 and 1964. As the baby boom population gets older, the younger generations will have the task of providing social services and support for the aging baby boomers. Countries with a high proportion of older people have fewer women of childbearing age, which means that the population in these countries tends to grow very slowly, if at all. Countries with high proportions of women of childbearing age, on the other hand, will tend to have a fast growing population.

Age–sex pyramids are graphs that illustrate important aspects of populations. They illustrate the composition of a population by age and sex. The horizontal bars represent different age groups, or **cohorts**, while the two sides of the pyramid represent the numbers or percentages of males and females in a population (see Figures 3.8, 3.9, and 3.11 in the textbook for examples of age–sex pyramids). Age–sex pyramids allow population geographers to identify changes in the age and sex compositions of a population. Age–sex pyramids can also be used to assess the future population growth of a country. An age–sex pyramid with a very narrow top and a wide bottom indicates a population that is growing rapidly. An age–sex pyramid with straight, even sides indicates a population that is stable and not growing at all. Age–sex pyramids for retirement communities, such as Sun City, Arizona, where most of the population is age 60 or older, look more like tornadoes than pyramids. Compare the age–sex pyramids in Figure 3.8 of the textbook to see the relationship between pyramid shape and population growth.

Demographers often divide the population reflected in an age–sex pyramid into three groups or cohorts. The **youth cohort** consists of people who are under 15 years of age, and are considered too young to be fully active in the labor force; that is, most of the people in this cohort do not hold full-time jobs. The **middle cohort** consists of people who are between the ages of 15 and 64, and who constitute the majority of the economically active, or working, population. Today's baby boomers constitute the middle range of this cohort. The **old-age cohort** consists of people who are 65 years of age or older, and who are often retired and considered to be beyond their economically active years. By dividing the total population into these three cohorts, two of which are not economically active, it is possible to compare the number of economically active people with those that are not. The result is the **dependency ratio**, which measures the dependency of the young and old on the middle, economically active sector of the population.

Population Dynamics and Processes

Population growth and change can be understood by looking at fertility, or birth rates, and mortality, or death rates. The number of people being born and dying in a population will affect its total size and rate of growth. The **crude birthrate (CBR)** is the total number of live births in a year for every thousand people in a population. The crude birthrate, like crude density, measures an average for a total population, and does not reflect variations within that population. Other measures may give a better idea of population growth potential. The **total fertility rate (TFR)** is a measure of the average number of children that a woman will have during her childbearing years, generally from ages 15 through 49. The total fertility rate is generally a more useful measure because it allows demographers to predict birthrates over time. A population with a total fertility rate of two represents replacement level fertility, because the two children born to that woman will replace the two parents. **Doubling time** is another measure of population increase. Doubling time refers to the number of years it will take for a population to grow to twice its size. At the present time, the doubling time for the world's population is about 40 years, which means that, at the current growth rate of 1.8 percent, the population of the world will be twice what it is at present in about 40 years from now.

Another factor influencing population is the **crude death rate (CDR)**, which is the ratio between the total number of deaths in one year for every thousand people in the population. Crude death rates—like crude birth rates—often reflect levels of economic development, and countries with low birthrates usually have low death rates. Other influences on mortality, or death rates, include the availability of health care, social class, occupation, and even place of residence.

In most places, the birthrate is higher than the death rate. This results in population growth, called **natural increase**. In some cases the number of deaths may be greater than the number of births, and this is called **natural decrease**. Death rates can be measured for certain cohorts of the population as well as for the population as a whole. The **infant mortality rate** is one such measure, and refers to the annual number of deaths of infants under one year of age compared to the total number of live births for that same year. Infant mortality rates generally reflect the level of a country's health care system as well as access to health care and nutrition. **Life expectancy** is another important indicator. Life expectancy is the number of years that an infant newborn can expect to live. This indicator, like the infant mortality rate above, varies not only from country to country but also within countries, often reflecting economic levels and access to health care and other resources. Life expectancy can be influenced by epidemics, which can quickly reduce population numbers. An epidemic having a devastating effect today is AIDS (Acquired Immunodeficiency Syndrome), which is not only an epidemic in the United States but also in many African countries.

Many demographers believe that fertility and mortality rates are connected to levels of economic development, arguing that the transformations associated with industrialization and urbanization influence population patterns in distinctive ways. The **demographic transition** is the movement from high birth and death rates to low birth and death rates. A model of this transition, the *demographic transition model*, shows four stages in the transformation. In the *first stage* of the demographic transition model, which represents pre-industrial societies, both birth and death rates are high. In the *second stage*, improvements in health care reduce the death rate, while the birthrate remains high because people in the society are still accustomed to having large numbers of children. In the *third stage*, the population becomes more urban, and family size decreases because children are not needed for agricultural work and because children in urban societies become more expensive to raise, and this lowers the birthrate. In the *fourth and final stage*, both birth and death rates are low, reflecting both technological improvements and changed social attitudes towards large families. This model is clearly illustrated in Figure 3.18 of the textbook. Demographers who use this model suggest that every country passes through similar stages, but at different periods in world history. Thus countries like the United States and Sweden have reached stage four, while many countries in Africa, for example, are still in stages two and three.

Population Movement and Migration

Birth and death are two factors influencing population change. A third major factor is the movement of people from one place to another, or migration. **Migration** is a long-distance move to a new location, for permanent or temporary residence. **Mobility** refers to the ability to move. Migration *from* a particular location is **emigration**, while migration *to* a particular location is **immigration**. Both of these factors can influence the size of a population. Demographers have developed ways of calculating migration rates, including **gross migration**, which is the total number of migrants moving into and out of a place or region, and **net migration**, which is the gain or loss in population in a particular area as a result of migration. Demographers have also looked into the factors responsible for migration, which can be divided into push factors and pull factors. **Push**

factors influence a person's movement *from* a particular place, and may include things like war, economic problems, and environmental and climatic factors. Unemployment and cold winters are both examples of push factors. **Pull factors** influence a person's movement *to* a particular location; they are forces of attraction that encourage people to move into a new region. Pull factors may include such things as good job opportunities or a good climate.

Migration to a new area may be either **voluntary**, in which the individual chooses to migrate, or **forced**, in which the migration is involuntary. Many people migrate voluntarily because of the reasons noted above and in the textbook, often because of better economic opportunities in the new country or region. You can get a better idea of why people move and how they adapt to their new home by taking a look at the Smithsonian Institution's website—see the textbook's website for the link. Figure 3.20 in the textbook illustrates global voluntary migration patterns for 2000–2005. The United States received many voluntary migrants throughout its history, and continues to do so. Migrants who take up jobs temporarily with the intention of someday returning to their own country are known as **guest workers**. Guest workers are usually attracted to job opportunities in other countries in which they can expect to be more economically successful than they could be in their home country. The demand for guest workers results from labor shortages in the countries receiving the workers. International forced migration results from factors beyond the migrants' control. The African slave trade is a good example of international forced migration. About 10 million Africans were brought against their will from Africa to the Americas and normally were unable to return to Africa. Internal wars can also result in international involuntary migration, with refugees fleeing the devastation of war in their home countries. Recent conflicts in such places as the former Yugoslavia and in Rwanda and Burundi have resulted in large numbers of refugees.

People also migrate within a country. The United States is a good example of internal voluntary migration. Population migrations within the United States have taken place in three waves. The *first wave* began in the middle of the 19th century and in some respects continues into the present. This wave was characterized by a population shift from rural to urban communities and by population movement from the East Coast towards the interior of the country. The *second wave* began in the 1940s and continued into the 1970s, and consisted of the massive movement of mainly African-Americans out of the rural South and into large cities in the South, North, and West. The *third wave* began in the 1940s and continues into the present, and is characterized by the migration of people from the Northeast and Midwest, sometimes known as the Snowbelt, Frostbelt, or Rustbelt, towards the warmer areas of the South and West, or Sunbelt. Florida and Arizona have been especially attractive to northern migrants and are among the fastest growing states in the country. Figure 3.24 in the textbook shows the changing population center of the United States, which has moved progressively towards the west and south, reflecting migration patterns in these two directions.

The American economy in the 20th century has largely shifted from reliance on the production of goods to the production of services. This economic change is reflected in population movements, as people leave the industrial cities of the Northeast and Midwest and move towards the South and West, where many high-tech and service industries are located. People have also moved out of the inner cities and into suburbs, in a process known as **suburbanization**. The growth of the suburbs relative to cities has resulted in many lifestyle changes, notably the dependence on the automobile as the principal form of transportation. Other countries have also experienced internal voluntary migration. In many countries of the periphery, migration has been largely from rural to urban areas, so that now many of the world's largest cities are located in countries of the periphery, including such places as Mexico City, São Paulo, Kolkata (Calcutta), and Cairo. Many of these migrants live in shantytowns on the fringes of the cities, and many city governments are unable to deal with the increased demands for services such as fresh water, sanitation, and education. Some governments have encouraged migration from cities to rural areas, sometimes by providing incentives and sometimes by making rural areas more accessible, through the building of new roads. The Brazilian government, for example, has encouraged migration to the Amazon region by building new roads into the jungles. Sometimes internal migration can be forced. The resettlement of Native American populations in the United States is an example of internal forced migration. In the "Trail of Tears" of the 1830s, nearly the entire Cherokee Nation was forced to move from their homeland in north Georgia to what is now Oklahoma, a completely different environment. In today's world, **eco-migration**, or migration stemming from environmental problems and the degradation of land and essential natural resources, has become a common form of internal forced migration.

Population Debates and Policies

How many people can the Earth support? This is a question that population experts have been asking for several centuries, and still debate today. An early and significant study of the relationship between population and resources was *An Essay on the Principle of Population*, published in 1798 by Thomas Malthus. Malthus argued that population grew at a faster rate than the food and resources needed to support it, and concluded that the world population would eventually outgrow its food supply. Though Malthus's dire predictions have not come true—so far technological improvements have allowed the food supply to keep up with population growth—many population experts, called Neomalthusians, argue that Malthus was essentially correct in that at some future point a population crisis will strike. Neomalthusians believe that there are natural limitations on the world's population, whereas many critics of the Neomalthusians argue that technology has and will continue to find solutions to population problems. You can find more data about how population changes will affect the United States by going to the textbook's website and following the link to the Why Population Matters site.

The idea that there are too many people in the world, or in a country or region, is known as *overpopulation*. The problem with this concept is that there is no objective measure of what "too many" really means. Many geographers argue that overpopulation may result from too many people compared to available food, water, and other natural resources, but even this concept is problematic. Japan, for example, has a large population and very few natural resources—food and many raw materials must be imported—yet we do not normally consider Japan to be overpopulated. Japan's economic wealth allows it to purchase needed food and resources while maintaining a large population. The question of overpopulation is largely a subjective one, with no clear right or wrong position. The perception that parts of the world are, or could become, overpopulated has led to the development of national and international **population**

policies and programs. Many of these policies are directed at lowering birth rates, such as family-planning programs that limit the number of children a couple may have.

Population is not growing at an even rate across the world. In the 21st century, geographers expect most population growth to take place in Africa, Asia, and Latin America, while Europe and North America will experience low or zero population growth. Countries with large populations, such as China and India, have been especially concerned about reducing population growth. China has instituted a one child per family policy, in which couples are permitted to only have one child. If additional children are born, the parents may face fines as well as denial of social services for the second child. India has used incentives, such as offering free contraceptives and family-planning counseling, in order to encourage couples to have fewer children. These policies have been somewhat successful in reducing population growth. Demographers believe that there is a strong link between birth rates and the status of women in a society. The more education a woman has, the less likely she is to have many children. Countries with more women in the labor force and that recognize greater rights for women also tend to have lower birth rates.

Birthrates, death rates, and migration are the central factors in population growth and change. Within the world-system, there are significant differences in birth and death rates among countries. Migration of populations has been especially important in shaping the present global system.

Review Questions

Some of the answers to these questions may be found only in the textbook, and not in the study guide.

1. _____ is the ratio between the number of agriculturalists per unit of arable land in a given unit of area.

2. True or False: A cohort is a group of individuals who share a common temporal demographic experience.

3. Information about births, deaths, marriages, divorces, and the incidence of certain infectious diseases is known as:
   a) census
   b) vital records
   c) medical geography
   d) demographic transition

4. True or False: The crude birthrate (CBR) is the ratio of the number of live births in a single year for every thousand people in the population.

5. _____ is the ratio between the number of deaths in a single year for every thousand people in the population.

6. True or False: Demographic transition is the total number of people divided by total land area.

7. Replacement of high birth and death rates by low birth and death rates is known as _____.

8. The study of the characteristics of human populations is:
   a) cartography
   b) regional geography
   c) population policy
   d) demography

9. _____ is a measure of the economic impact of the young and old on the more economically productive members of the population.

10. True or False: Doubling time is the measure of how long it will take the population of an area to grow to twice its current size.

11. Crude density is the total number of _____ divided by the total _____.

12. Population movement caused by the degradation of land and essential natural resources is:
    a) eco-migration
    b) internal migration
    c) demographic transition
    d) suburbanization

13. True or False: emigration is a move *from* a particular location.

14. _____ are events and conditions that impel an individual to move *from* a location.

15. Movement by an individual against his or her will is known as _____.

16. _____ is the practice of assessing the location and composition of particular populations.

17. Which of these countries would have an age–sex pyramid that is most pyramidal in shape: Kenya, United States, Denmark? _____. Why?
_____
_____

18. A representation of the population based on its composition according to age and sex is called an _____.

19. The total number of migrants moving into and out of a place, region, or country is known as:

      a) internal migration

      b) international migration

      c) gross migration

      d) arithmetic density

20. _____ are individuals who migrate temporarily to take up jobs in other countries.

21. Approximately what percentage of the world's population lives in Asia? _____.

22. A move *to* another location is called:

      a) emigration

      b) immigration

      c) gross migration

      d) international migration

23. The annual number of deaths of infants under one year of age compared to the total number of live births for that same year is the _____.

24. True or False: International migration is a move within a particular country or region.

25. _____ is the average number of years an infant newborn can expect to live.

26. The sub-area of geography that specializes in understanding the spatial aspects of health and illness is _____.

27. The population of individuals born between the years 1946 and 1964 is known as:

    a) Generation X

    b) baby boom

    c) original generation

    d) old people

28. Members of the population aged 15 to 64 who are considered economically active and productive are called the _____.

29. True or False: Mobility is the ability to move, either permanently or temporarily.

30. _____ is the difference between the CBR and CDR, which is the surplus of births over deaths.

31. A move from one country to another is called _____.

32. Thomas Malthus argued that population size increases more rapidly than does the _____.

33. The gain or loss in the total population of a particular area as a result of migration is known as:

    a) net migration

    b) emigration

    c) immigration

    d) gross migration

34. The ratio between the total population and the amount of land under cultivation in a given unit of area is _____.

35. The most widely used method for assessing the state of a population is the _____.

48   Chapter 3

36. Members of the population aged 65 years of age and older who are considered beyond their economically active and productive years constitute the:

   a) youth cohort

   b) middle cohort

   c) old-age cohort

   d) baby boom

37. True or False: Pull factors are forces of attraction that influence migrants to move *to* a particular location.

38. The growth of population along the fringes of large metropolitan areas is known as _____.

39. True or False: Natural increase is the difference between CDR and CBR, which is the deficit of births relative to deaths.

40. The average number of children a woman will have throughout the years that demographers have identified as her child-bearing years, approximately ages 15 through 49, is known as _____.

41. _____ is movement by an individual based on choice.

42. The count of the number of people in a country, region, or city is a:

   a) census

   b) crude birth rate

   c) middle cohort

   d) dependency ratio

Geographies of Population 49

Conceptual Questions

1. Using census data for your community, construct an age–sex pyramid [you can get this data from the United States Census Bureau (www.census.gov)]. What does this pyramid tell you about your community?

_____
_____
_____
_____
_____
_____
_____
_____
_____
_____
_____

Population pyramids for three St. Louis County, Minnesota census tracts, 2000

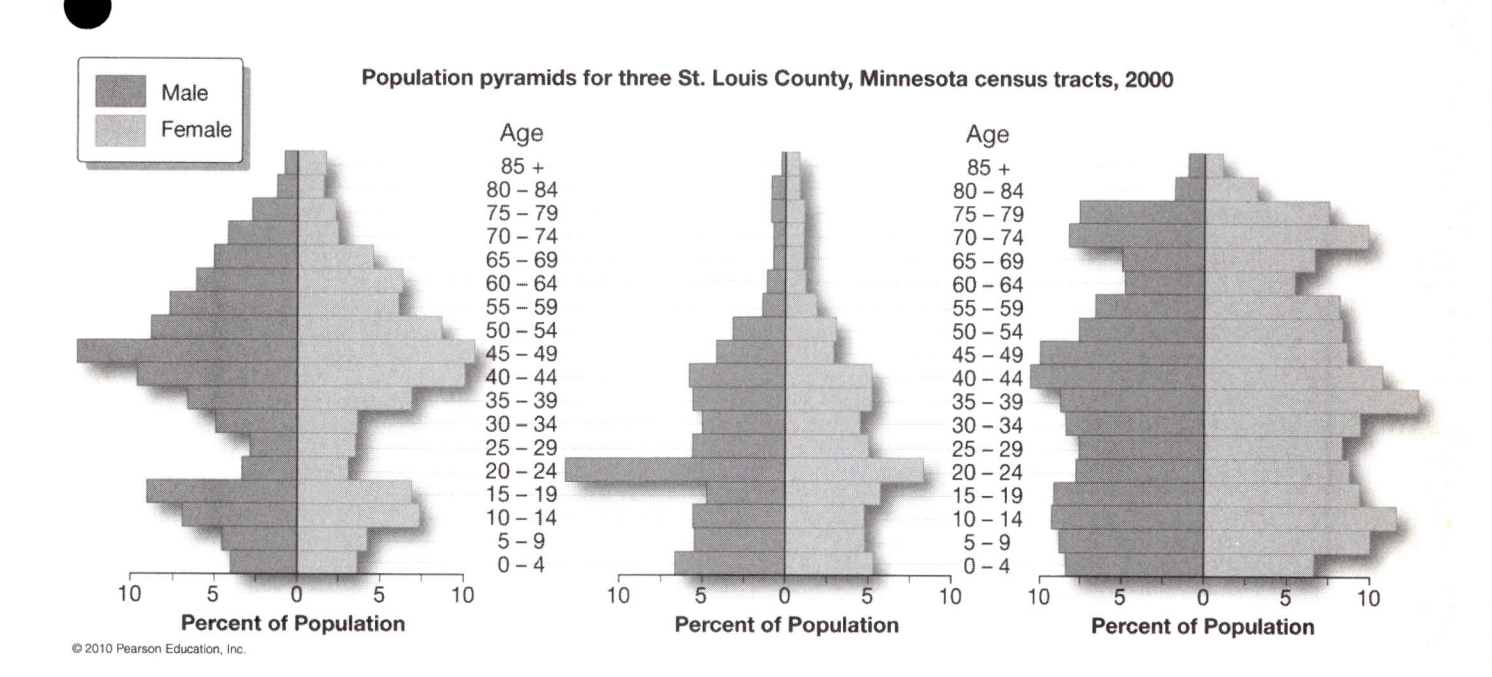

Figure 3.11

© 2010 Pearson Education, Inc.

50   Chapter 3

2. Look at the three age–sex pyramids in Figure 3.11, which are from three different census tracts in the St. Louis County, Minnesota area. What conclusions can you draw about the income level in each census tract? What other points do these age–sex pyramids illustrate?

3. What are the basic limitations of census data? How do you think that gathering census data could be improved?

4. How is the world's population distributed? What factors account for the concentration of population in some areas, while other areas have almost no population?

5. What is meant by the term "baby boom"? How is this generation likely to affect life in the United States in the next 20 years?

6. Outline the demographic transition model. Can you give examples of countries within each of the four stages?

7. Using your own family as an example, describe where you or your ancestors migrated to and from. What factors accounted for this migration?

8. Do you think the world, or some places in the world, is overpopulated today? Why or why not?

4

Nature and Society

Chapter Objectives

The objectives of this chapter are to:

1. Understand nature as a concept
2. Investigate Earth's transformation by ancient humans
3. Explore European expansion and globalization
4. Examine recent environmental change through human action

Chapter Notes

Nature as a Concept

Concern for the environment and the relationship between society and nature are two issues gaining importance today. These issues are gaining importance because of the persistence of environmental problems. In the past, technology appeared to be capable of solving most if not all environmental problems, but that does not seem to be the case today. Much of the research on society–nature relations is concerned with the failure of technology to solve current problems. Before exploring the relationship between nature, society, and technology, we need to clearly define these terms. **Nature** is defined as not only the physical surroundings, but also as a social creation—attitudes towards nature. Human beings are part of this concept of nature. **Society** is the sum of the inventions, institutions, and relationships created and reproduced by human beings across particular places and times. Society is connected to nature because nature is one of those relationships—the relationship between people and the physical world. **Technology** mediates between society and nature, and is defined as physical objects or artifacts, activities or processes, and knowledge or know-how.

Different societies will have different impacts on their environments. Generally, wealthier, more populous, and technologically complex societies have a greater impact on the environment. This idea is expressed in the formula $I=PAT$, where I is impact on the Earth's resources, P is population, A is affluence, and T is technology. The three factors multiplied together give the impact on Earth's resources. Technology affects the environment in three ways: through the harvesting of resources; through the emission of wastes in the manufacture of goods and services; and through the emission of waste in the consumption of goods and services. Technology can be both a solution and a problem. For example, nuclear energy is in some ways cleaner and more efficient than burning fossil fuels, yet it can also lead to harmful effects such as the release of radiation into the atmosphere. **Cultural ecology** is the study of how human society has adapted to environmental challenges such as aridity and steep slopes through technologies such as irrigation and terracing and the organization of people to construct and maintain these systems. It is most closely associated with the work of Carl Sauer, a former professor at the University of California at Berkeley.

Though the European and Euro-American perspective on nature is dominated by its Judeo-Christian roots, there have been many differing minority viewpoints within the larger tradition. Henry David Thoreau (1817–1862), an American naturalist and writer, wrote several books—including *Walden*, his best known—in which he challenged the dominant view of nature as an exploitable resource, arguing instead for greater respect for nature. His experiment living at Walden Pond outside Boston showed that even Americans of the industrializing 19th century could find peace, solace, and inspiration in nature. Thoreau was part of a larger movement called **Romanticism**, a philosophy originating in Europe that emphasized the interconnections between society and nature. One of the best-known books of the Romantic era was Mary Shelly's *Frankenstein*, which warns of the dangers of science and technology when they get out of control. A distinctly American version of Romanticism was **trancendentalism**, a mystical philosophy in which a person attempts to rise above the limitations of the body and achieve a greater spiritual awareness of nature.

The early 20th century saw the rise of two important environmental positions known as conservation and preservation. **Conservation** was associated with such figures as President Theodore Roosevelt and his Chief Forester Gifford Pinchot. Conservation is the view that natural resources should be conserved and used wisely, so that they are not damaged, wasted, or destroyed. The conservation view guided the development of the United States Forest Service and its "multiple use" management perspective. **Preservation** is a position arguing that certain habitats, species, and resources should be off-limits to human use; that is, that they should be preserved and not used or exploited. One of the leading lights of preservationism was John Muir, who was one of the founders of the Sierra Club in the early 20th century. John Muir was also an explorer, naturalist, and writer, and was the key figure in preserving Yosemite Valley in California as a national park. Environmental groups today usually hold some combination of the above two positions. The Sierra Club, despite its founding by John Muir, takes a balanced approach between conservation and preservation, while more radical groups like Earth First! are clearly preservation-oriented, often using eco-terrorist tactics, or "monkey-wrenching," to protect the environment from development. The conflicts between different attitudes to the environment are apparent in the controversy over reintroducing wolves to parts of the American Southwest. You can follow this issue by checking out the Reintroduction of the Wolf into the Southwest U.S. website—follow the link through the textbook's website.

Environmental ethics is a philosophical perspective on nature that prescribes moral principles as guidance for our treatment of it. The perspective of environmental ethics argues that people should treat nature in the same way as they treat each other—in other words, that nature also has rights. Thus, if human beings have the right not to be abused or exploited, then nature should also have this right. A similar perspective on society–nature relations is **ecofeminism**. Ecofeminists believe that patriarchy—the valuing of men's ideas and ways more highly than women's—is the cause of present-day environmental problems. Ecofeminists argue that women—as opposed to men—have a different perspective on nature, one that is more nourishing and respectful. Chief among the objectives of ecofeminists is increasing the value of cultural and biological diversity, away from the male-controlled society of the present. **Deep ecology** is a third radical perspective on the environment. Deep ecology is similar to some Asian perspectives on nature in that it does not see human beings as above nature. Deep ecologists call for self-realization and biospherical egalitarianism, in which all forms and aspects of nature deserve equal consideration and treatment. A major factor in the perspective on the environment dominating European and Euro-American culture today is science. Science takes a non-spiritual, mechanistic view of the environment, attempting to remove values from it. Within science, nature is viewed as operating according to certain fixed laws, which human beings can discover in order to improve their own livelihoods. The rise of science largely displaced earlier views of the Earth as a living organism, replacing it with a concept of nature as something like a giant machine.

The Transformation of Earth by Ancient Humans

Prehistoric peoples altered their environment even though they lacked machines or elaborate tools. In the **paleolithic period**, about 1.5 million years ago, when chipped stone tools were first used, people lived in small groups and subsisted on hunting and gathering. These people used fire in various ways, leading to changes in vegetation. Paleolithic people may also have had an impact on the extinction of prehistoric mammals through over-hunting. Further impacts on the environment came with agriculture, which was practiced by neolithic, or late stone age, peoples about 10,000 years ago. The early domestication of plants and animals began a process that continues into the present—that of genetically changing plants and animals in order to achieve desired characteristics. One impact of agriculture and domestication was the alteration of **ecosystems**, or communities of species interacting with each other and the surrounding physical environment, into new ecosystems with fewer varieties of plants and animals. Probably the most significant aspect of plant and animal domestication was that it enabled a surplus to be produced. The production of a surplus meant that some people could live off the surplus and did not have to gather food. This led to changes in societies, with the development of political and religious elites and of craftspeople and artisans. The development of agricultural tools and methods of irrigation also contributed to changes in the environment. Some early civilizations, such as those in Mesopotamia in what is now Iraq, may have mismanaged their environments in ways that led to the collapse of those civilizations. **Deforestation**, the removal of trees from a forested area without adequate replanting, and **siltation**, the build up of sand and clay in a waterway, resulted from agricultural practices yet made continued agriculture increasingly difficult. Though not on the scale seen at present, environmental damage was a feature of ancient civilizations as well as modern ones.

European Expansion and Globalization

Europe is an example of a society with distinctive attitudes towards the environment—attitudes that contributed to European global expansion and technological advancement. The European perspective centered around science, the Judeo-Christian religious tradition, and a capitalist political and economic system. At first, European expansion was internal—European societies expanded within Europe itself. Population was growing relatively rapidly in the Middle Ages, and this increasing population needed adequate food resources. More forests were cut down and marshes and fens drained to make room for agriculture, more animals were killed for food, and minerals and other resources were exploited. This internal expansion was largely completed by the year 1300. Beginning in the 15th century, European expansion was largely external. Europeans began to explore and settle—through the process of colonialism—other parts of the world. Within several centuries, Europeans politically and economically controlled many parts of the world. European customs and values, including attitudes toward the environment, were introduced as part of the process of colonialism.

European contact with and expansion into the New World devastated local populations, mainly due to the introduction of new diseases to which the residents of the New World had no natural immunity. This lack of immunity, indicating that the oldest member of the group had no previous exposure to the disease in question, is known as a **virgin soil epidemic**. The interaction between the Old and New Worlds beginning at the time of Columbus is known as the **Columbian Exchange**. In addition to new plants and animals introduced into the New World, Europeans also brought many new diseases, including smallpox, measles, chicken pox, whooping cough, typhus, typhoid fever, bubonic plague, cholera, scarlet fever, malaria, yellow fever, diphtheria, influenza, and many others. None of these diseases existed in the New World before European contact. One of the few diseases that traveled in the other direction, from the New World to the Old, was syphilis. The devastating effect of the introduction of new diseases into the New World weakened the indigenous populations and made the European work of conquest easier. The effects of these new diseases were something like genocide, and are known as **demographic collapse**. Environmental impacts stemming from the death of large numbers of people included abandonment of settlements and agricultural land and the reversion of this land to jungle and forest.

Europeans introduced new plants and animals into the New World. This process is known as **ecological imperialism**, which means that exotic, non-native plants and animals are introduced into new ecosystems. Europeans introduced new plants, such as wheat, sugarcane, citrus fruits, coffee, and tumbleweeds, and new animals, such as cattle, horses, sheep, and goats, into the New World. Europeans also brought New World plants and animals—such as corn, potatoes, tobacco, cocoa, tomatoes, and turkeys—back to the Old World. The Columbian Exchange altered the agricultural practices and eating habits of both Old and New Worlds. Tomatoes and potatoes did not exist in Europe before European expansion, and Italian and German cuisine were something quite different from what they're like today (the textbook's website has a good link to the World Geography of Potatoes website, which follows the odyssey of this important root). Likewise, the European introduction of the horse changed social and hunting patterns among many Native American peoples. The introduction of the horse and other work animals such as the ox increased the animal power available for work, for pulling plows and for transportation. The use of irrigation and intensive agriculture by such peoples as the Aztecs and the Incas resulted in some deforestation and increased soil salinity.

Human Action and Recent Environmental Change

Industrialization and urbanization are the two processes that have most dramatically revolutionized human life, and, incidentally, have had the greatest impact on ecological change. At one time, these impacts were largely local or regional, but today they can affect the entire planet. The most important technological breakthrough of the Industrial Revolution was the discovery and utilization of fossil fuels such as coal, oil, and natural gas. The increased power that these fuels provided allowed for greater industrial expansion. Today the world relies on fossil fuels for the majority of its energy needs. The production and consumption of energy resources varies greatly by region. Core countries tend to use more energy per person than do peripheral countries, as shown in Figure 4.18 in the textbook. The production and consumption of energy resources also has significant impacts on the physical landscape. For example, mining coal or drilling for oil may result in damage to plants and animals and in erosion and water pollution, while the burning of coal and oil emits polluting substances into the atmosphere and may contribute to climate change. Intermediary processes, such as transportation of fossil fuels, can also have environmental impacts, from the construction of roads and pipelines into remote wilderness areas to oil spills on both sea and land. Natural gas, though often viewed as cleaner than either coal or oil, can also have adverse environmental impacts, and the transportation of natural gas in liquefied form is particularly dangerous.

In the second half of the 20th century, many people saw nuclear energy as a viable, clean, and safe alternative to fossil fuels. It was only after such major nuclear disasters as Windscale in England and Three Mile Island in the United States that many people began to question nuclear power. The meltdown of the Chernobyl nuclear reactor in 1986, in what was then the Soviet Union, further indicated the disastrous potential of nuclear power. Many core countries have started to reduce their dependence on nuclear power, while many periphery countries have increased it. Many people in peripheral countries depend on wood for fuel, often because they cannot afford or lack access to fossil fuels. As the population of the periphery grows, so does the demand for wood, resulting in deforestation, erosion, and air pollution. Hydroelectric power is another important source of energy. Though not leading to air pollution directly, as does the burning of fossil fuels, hydroelectric power can have major negative environmental impacts. The construction of large dams to generate power can flood large areas, as, for example, the Glen Canyon Dam on the Colorado River in the United States and the various dams comprising the James Bay project in northern Québec, Canada, have done. Impoundment of water not only causes flooding, which destroys many living things, but also changes the flow and quality of water in rivers. Water near the mouth of the Colorado River, for example, is so salty that it cannot be used for agriculture or for drinking. Impounded, still water can also serve as a breeding ground for mosquitoes and other insects, and reservoirs constructed in the Amazon region of Brazil have increased the numbers of malaria-carrying mosquitoes in the region. **Acid rain** is a problem associated with air pollution. Increased levels of acidic substances such as sulfur dioxide, nitrogen oxides, and hydrocarbons, released into the atmosphere through motor vehicle exhausts, industrial processes, and power generation, can result in the deposition of these substances on the ground or in water through precipitation in the form of rain or snow. Acidification can kill plants and animals on land and water. Alternatives to the above energy sources are available, and include geothermal, wind, solar, and tidal power. Energy conservation is perhaps the best alternative to the continued substantial use of energy resources.

    Pollution is not the only result of increased demands for energy. The way land is used can also be affected. Land may be divided into five categories: forest, cultivated land, grassland, wetland, and areas of settlement. Land use change can take place in one of two ways, by conversion, in which land changes from one use to another, or modification, in which the existing use is altered. Human actions have had a great impact on forests—forests have been greatly reduced in area throughout much of human history. About 3 million square miles of forest have been cut down since preagricultural times. Much of the concern today is over the destruction of the tropical rainforests in South America, Africa, and Asia. Not only are trees being lost, but ecosystems are being changed, and many species are being lost. See also the World Resources Institute's Forest Frontiers Initiative website by following the link from the textbook's website. Cultivating land also has environmental impacts. The amount of land under cultivation in the world has steadily increased, as population has grown and demands for food increase.

Grasslands are also used productively, especially for the grazing of animals. Overgrazing can lead to **desertification**, which refers to the degradation of land cover and damage to the soil and water in grasslands and arid and semiarid lands. Lands become more desertlike, reducing the potential for productive use and making it difficult to restore the original ecosystem. Land on the margins of existing deserts—such as in the American Southwest and the Sahel region south of the Sahara Desert in Africa—is especially prone to potential desertification. Wetlands include swamps (which contain trees), marshes (which do not have trees), and the shore areas of bodies of water. Many wetlands have been drained or filled in to increase dryland areas, especially for settlement and agriculture. For example, large parts of San Francisco Bay, in California, have been filled in to increase the area available for building and settlement. In addition to destroying the existing ecosystem, the filled-in areas are especially unstable in earthquakes, as the 1989 Loma Prieta earthquake in San Francisco demonstrated. Much of the worst damage in that earthquake was suffered by structures built on landfill.

At one time the environmental changes and impacts mentioned above were limited to a local or regional level. Today, the impacts are often global (see the United States Environmental Protection Agency's website for a series of air, land, and water maps—follow the link from the textbook's website). Radiation from nuclear accidents on one continent can be detected on others, and the burning of rainforests in Brazil can have an impact on the entire world's climate. The interrelationship and combination of political, economic, social, historical, and environmental problems at the world scale is known as **global change**. Environmental problems are not always felt equally by everyone. Often poorer people or poorer countries have a disproportionately higher share of suffering from environmental problems: toxic wastes are dumped in poorer neighborhoods, or mines are opened in peripheral countries and not in wealthier core countries. Some information about these issues in the city of Los Angeles can be gleaned from the Los Angeles: Revisiting Four Ecologies website that is linked to the textbook's website. **Environmental justice** is a movement that seeks to redress this perceived imbalance between wealth and share of suffering from environmental problems. Activists calling for environmental justice see a need for a more equitable distribution of the world's resources.

The Globalization of the Environment

Globalization has resulted in an increasingly shrinking world, in terms of increased flows and connections between people and societies. Political action has also become global. The number of environmental organizations with a global membership and mandate has also increased as environmental problems are recognized as being global in nature. Globalization has also resulted in an increase in the number of international environmental agreements. Many of these agreements have centered around the concept of *sustainable development*, a concept that has various definitions but generally is concerned with recognizing the role of the environment, and impacts on it, in the economic development process.

Review Questions

Some of the answers to these questions may be found only in the textbook, and not in the study guide.

1. The wet deposition of acids upon the earth created by the natural cleansing properties of the atmosphere is known as _____.

2. _____ is the build-up of sand and clay in a natural or artificial waterway.

3. The influential nature book *Walden*, authored by _____, regarded the natural world as an antidote to the negative effects of technology on the American landscape and the American character.

4. The _____ is a flaked, bifaced projectile whose length is more than twice its width.

5. The Kyoto Protocol marks the first time that attempts have been made to limit the amount of _____ generated by core countries.

6. The interaction between the Old World, originating with the voyages of Columbus, and the New World is known as:

   a) deep ecology

   b) Columbian exchange

   c) virgin soil epidemics

   d) Paleolithic period

7. True or False: Conservation is the view that natural resources should be used wisely, and that society's effects on the natural world should represent stewardship and not exploitation.

8. _____ was an American biologist who published the first piece of scientific research on the environmental effects of pesticides.

9. True or False: Conservation is the approach to nature revolving around two key components: self-realization and biospherical egalitarianism.

10. _____ is the removal of trees from a forested area without adequate replanting.

11. Technology affects the environment in what three ways?

_____

_____

_____

12. The degradation of land cover and damage to the soil and water in grasslands and arid and semi-arid lands is known as:

　　a) virgin soil epidemic

　　b) demographic collapse

　　c) deforestation

　　d) desertification

13. _____ and _____ are the two major processes that have revolutionized human life and effected far-reaching ecological changes.

14. The view that patriarchal ideology is at the center of our present environmental malaise is called:

　　a) deep ecology

　　b) political ecology

　　c) ecofeminism

　　d) romanticism

15. The introduction of exotic plants and animals into new ecosystems is known as:

　　a) demographic collapse

　　b) ecological imperialism

　　c) siltation

　　d) virgin soil epidemic

16. A community of different species interacting with each other and with the larger physical environment that surrounds it is an _____.

17. The movement that reflects a growing political consciousness largely among the world's poor that their immediate environs are far more toxic than wealthier neighborhoods is called:

　　a) ecofeminism

　　b) environmental ethics

c) environmental justice

d) preservation

18. The combination of political, economic, social, historical, and environmental problems at the world scale is referred to as _____.

19. The largest proportion of the world's current consumption of energy resources comes from _____.

20. True or False: Nature is a social creation as well as the physical universe that includes human beings.

21. _____ and _____ are the two ways that land-use change occurs.

22. The _____ is the period when chipped stone tools first began to be used.

23. In the formula $I=PAT$, what does each letter stand for?

_____

_____

24. The approach to nature advocating that certain habitats, species, and resources should remain off-limits to human use, regardless of whether the use maintains or depletes, is called _____.

25. _____ is the phenomenon of near genocide of native populations.

26. The philosophy that emphasizes interdependence between humankind and nature is known as:

a) romanticism

b) deep ecology

c) conservation

d) preservation

27. _____ is the sum of the inventions, institutions, and relationships created and reproduced by human beings across particular places and times.

28. _____ is a philosophical perspective on nature that prescribes moral principles as guidance for our treatment of it.

29. True or False: Society refers to physical objects or artifacts, activities or processes, and knowledge or know-how.

30. _____ is a philosophy in which a person attempts to rise above nature and the limitations of the body to the point where the spirit dominates the flesh.

31. The condition in which the population at risk has no natural immunity or previous exposure to the disease within the lifetime of the oldest member of the group is called:

a) transcendentalism

b) virgin soil epidemic

c) ecological imperialism

d) siltation

32. The merging of political economy with cultural ecology is called _____.

33. The interaction between the New World and the Old World—for example, the introduction of new diseases—is known as the _____.

Conceptual Questions

1. What impact did the romantic and transcendentalist movements have on European and American life? How are these impacts felt today?

68 Chapter 4

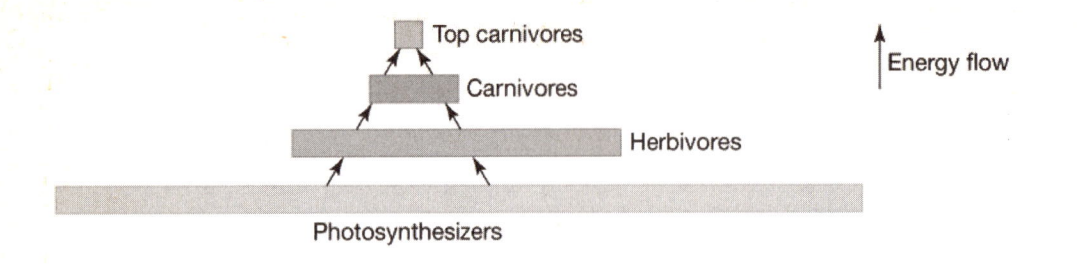

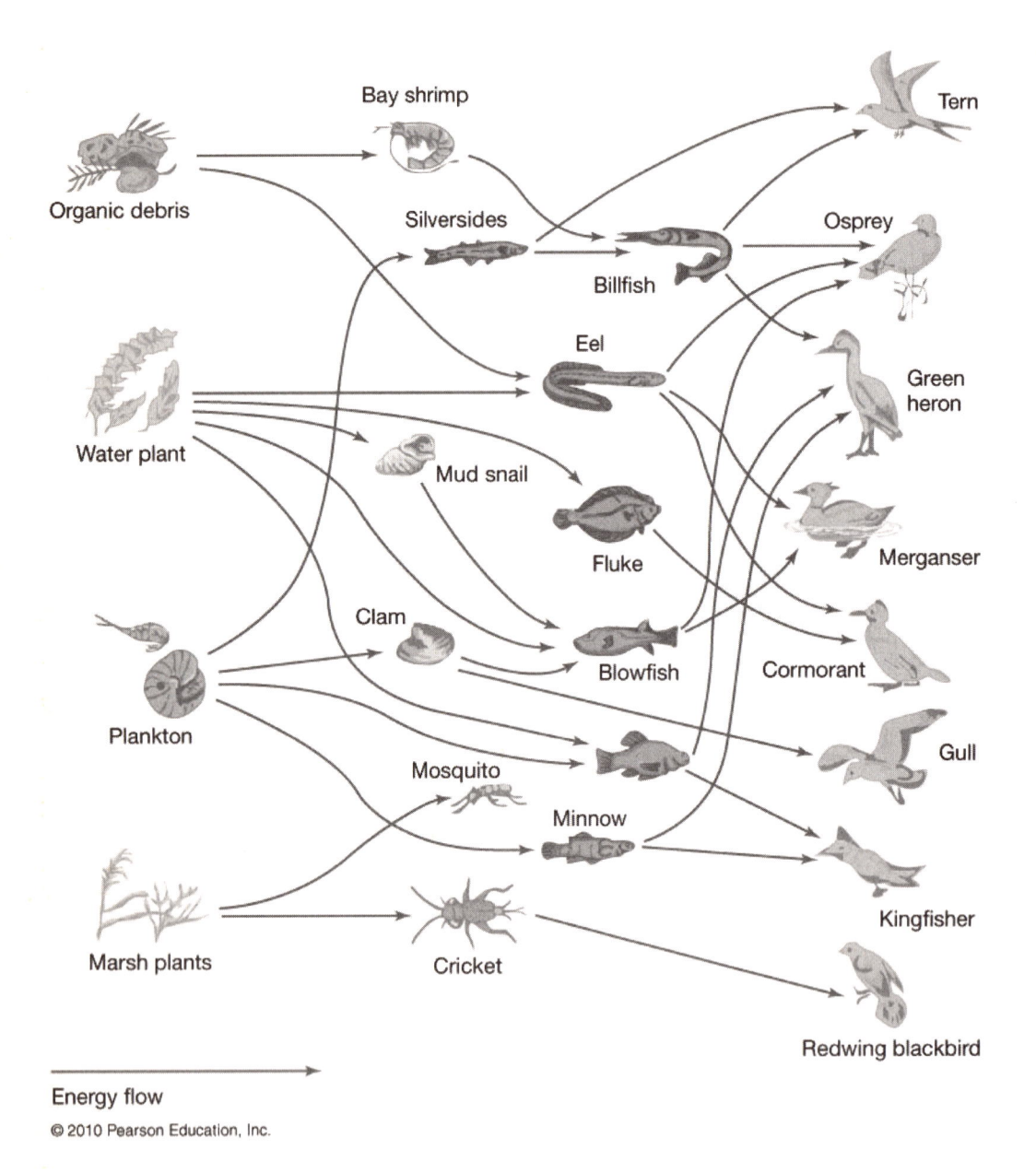

Figure 4.6

2. Figure 4.6 shows a generalized food chain. Describe the potential impacts on the food chain if: 1) clam were removed; 2) cormorant were removed; 3) plankton were removed; and 4) minnow were removed.

3. What are the basic differences between conservation and preservation? Which view do you share, and why?

4. Describe the major impacts that prehistoric peoples had on their environment. Why did these impacts occur?

5. What is meant by the Columbian Exchange? What things were exchanged, and why?

6. Identify the major sources of energy in the world today. What costs and benefits are associated with each of these sources?

7. Why has the amount of land covered by forests been steadily decreasing throughout human history? What environmental impacts has this deforestation had?

8. Look at Figure 4.23 and Figure 4.24, which illustrate the worldwide distribution of nuclear reactors and the global use of fuelwoods. What kinds of conclusions can you draw about the global distribution of wealth and access to energy from these figures? How do they relate to each other?

72 Chapter 4

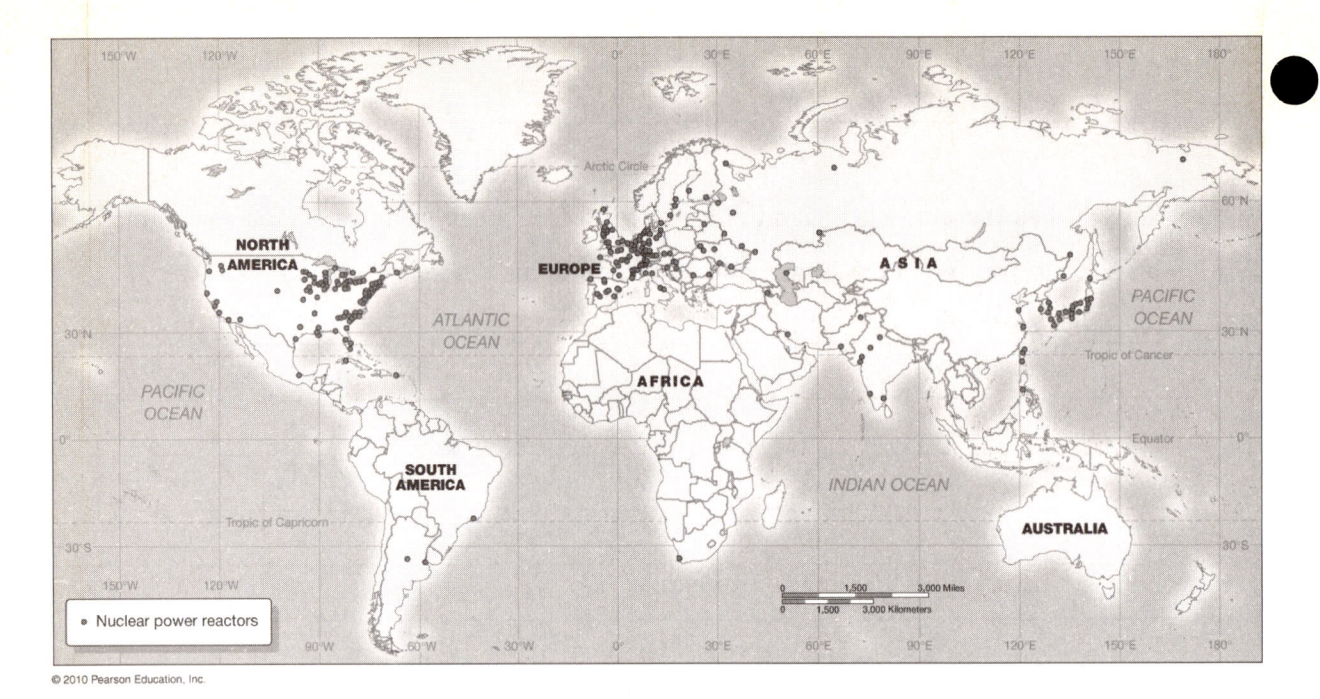

Figure 4.23

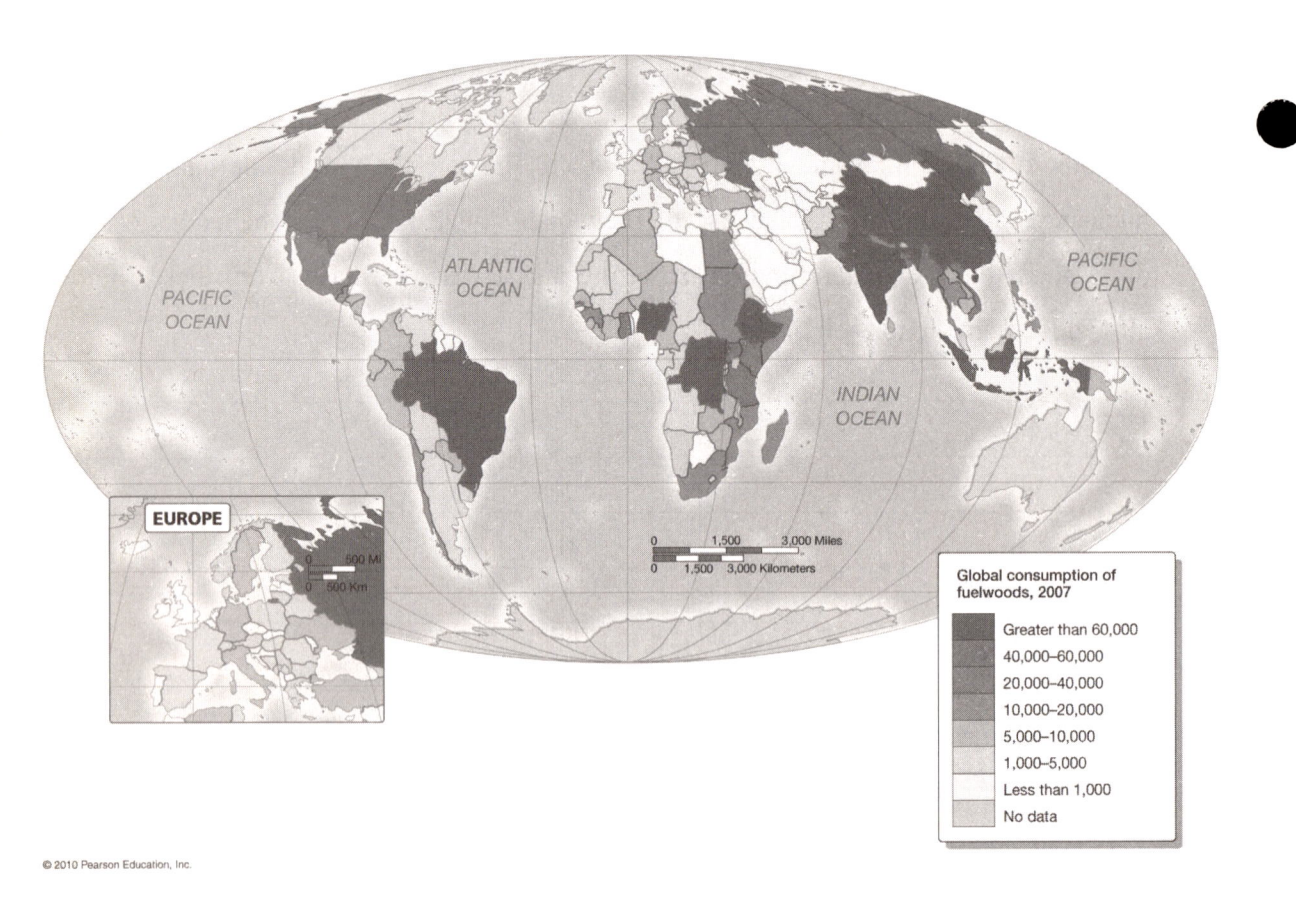

Figure 4.24

5

Cultural Geographies

Chapter Objectives

The objectives of this chapter are to:

1. Examine the cultural systems of religion and language as they are related to geography
2. Provide a foundation in the understanding of cultural nationalism
3. Survey culture and identity by examining our sexual geographies, ethnicity and the use of space, race and place, and gender
4. Explore the relationship between culture and the physical environment
5. Investigate the relationship between globalization and cultural change

Chapter Notes

Culture as a Geographical Process

Many social science disciplines study different aspects of culture. Geographers studying culture focus on how place and space shape culture, and on how culture shapes place and space. **Culture** can be defined in many different ways. In a broad sense, culture is a shared set of meanings that are lived through the material and symbolic practices of everyday life. These shared meanings can include values, beliefs, practices, and ideas about religion, language, family, gender, sexuality, and other important identities. Culture can be altered and changed by factors arising inside and outside the culture group. **Cultural geography** is a field within geography that focuses on how space, place, and landscape shape culture, and how culture shapes space, place, and landscape. Cultural geographers explore both globalizing trends as well as what makes culture distinctive in different localities.

Building Cultural Complexes

The interaction between people and their landscape changes the landscape. The outcome or result of this interaction between a human group and a natural environment is called a **cultural landscape**. Carl Sauer (1889–1975) was a geographer at the University of California at Berkeley who helped establish cultural geography as a distinct field. Sauer's work emphasized the interaction between people and their environment, and the distinction between a natural and a cultural landscape. Figure 5.4 in the textbook presents a brief outline of some of his main concepts. **Historical geography** is associated with Great Britain and especially with the British Geographer H.C. Darby (1909–1992). Darby analyzed archival records in order to understand how cultural landscapes evolve. ***Genre de vie***, a concept associated with the French geographer Paul Vidal de la Blache (1845–1919), refers to the lifestyles connected with particular places. Vidal de la Blache looked at how people adapted to changes in their lifestyles, changes that largely originated outside their locality.

Cultural traits are *specific* aspects of a given culture. For example, not eating beef is a cultural trait of Hindus, while tattooing one's body is a cultural trait of many Polynesian peoples. The combination of cultural traits is a **cultural complex**, which includes all the traits characteristic of a specific cultural group. A **cultural region** is the area in which a particular cultural system prevails. Cultural regions can be understood at many levels, involving all aspects of culture. For example, India would constitute a cultural region in which Hinduism is the dominant religion. This factor distinguishes India from its neighbors. Within India itself, one could find smaller cultural regions based on religion, for example, cultural regions where Islam or Jainism is dominant. Cultural regions may be based on factors other than religion. Within Europe, for example, one could speak of two cultural regions, northern and southern, based on whether beer or wine was the alcoholic beverage most commonly consumed.

Cultural Systems

Broader than the cultural complex concept is the concept of a **cultural system**, a collection of interacting elements that taken together shape a group's collective identity. Two important components of a cultural system are a group's religion and language. These two components may unite the smaller cultural complexes while still maintaining variation within them.

Religion is a belief system and a set of practices that recognizes the existence of a power higher than humans. Whether or not one practices or believes, religion may still shape social values and attitudes within a society. Europeans and North Americans who do not practice Christianity (or even any other religion) still have their values and lifestyles shaped in many ways, and to varying degrees, by Christianity. Christianity has shaped both European and North American political and economic systems, for example. Take a look at the Geography of Religion Basics and the Geography of Religion websites, linked through the textbook's website, for detailed information on religion and geography. The most important influence on religious change has been conversion—changing from one religion to another. Christianity has become the most widely practiced world religion largely through conversion, initiated by missionary activity. Many of the inhabitants of the New World, for example, were convinced or forced to give up their traditional religions and convert to Christianity. People also convert voluntarily if they feel that another religion best reflects their values and beliefs. **Diaspora** is a spatial dispersion of a previously homogeneous group.

Many of the world's religions have spread out from or moved away from their traditional areas of practice. Figure 5.10 in the textbook illustrates the current distribution of the world's major religions, though one should keep in mind that this map only indicates the most widely practiced religion in each area. Many of the world's religions have diffused or spread out from their places of origin. Figure 5.11 in the textbook shows the diffusion of four major religions: Christianity, Islam, Hinduism, and Buddhism. These four religions all originated in Asia, but two of them, Christianity and Islam, have become well-established outside that continent. Religious traditions can also combine—especially where different cultures mix—to create new, *syncretic* religions that fuse elements of each source religion. Voodoo and Santería in the Caribbean region are examples of syncretic religions that combine African and European religious traditions.

Language is another aspect of cultural systems. Language is the means of communicating ideas or feelings by means of a conventionalized system of signs, gestures, marks, or articulate vocal sounds. Language is symbolic, using commonly understood sounds or signs. Within each language, a number of **dialects**, or regional variations, may exist. Languages can be classified into families, branches, and groups. A **language family** is a set of languages related to a common ancestor language. Thus all the individual Indo-European languages, such as English, Spanish, Russian, and Hindi, are related to a common ancient proto-Indo-European language. A **language branch** is a set of languages that have a common origin but have split into individual languages. The Indo-European language family includes such branches as Slavic, Germanic, and Romance. A still smaller set is the **language group**, which consists of languages that are part of a language branch and have a similar grammar and vocabulary. Serbian and Slovenian are both languages in a common language group, the South Slavonic.

Identifying the source regions, or **cultural hearths**, of innovations, ideas, or ideologies is a traditional approach to cultural geography. Languages and religions, among other cultural features, can often trace their origins to a single source or hearth. Globalization also has an impact on cultural features such as languages. The variety of languages spoken within a single political unit may make governing difficult, and often one language will be given official status, sometimes to the detriment of the others. Some languages can replace others, so that now English has become the most important language of international commerce while many smaller indigenous languages, such as those of Africa and North America, are being lost. The dominance of the English language is examined in the Internet and the English Language website, linked to the textbook's website.

Islamic Cultural Nationalism

Cultural nationalism is an effort to protect regional and national cultures from the homogenizing impacts of globalization, especially the penetrating impacts of U.S.

culture. U.S. culture is increasingly dominant in the world today, reflected in such things as the dominance of the English language, U.S. media such as television and film, U.S. music, fashion, styles, and foods, among many other things. One can find McDonald's hamburgers, drink Coca-Cola, and listen to American rock music throughout the world.

Islam is the second most widely practiced religion in the world (after Christianity) and provides a good example of the geography of religion. An adherent to Islam is a **Muslim**, literally "one who submits." Islam recognizes the prophets of the Old and New Testaments of the Bible, but considers Muhammad to be the last prophet and God's messenger on Earth. The *Qur'an* (sometimes spelled *Koran*) is the principal holy book of Islam. The cultural hearth of Islam is Southwest Asia, the area known as the Middle East. Islam is still widely practiced in this region, but has diffused well beyond the original hearth. Figures 5.26 and 5.27 in the textbook depict the cultural hearth and diffusion of Islam throughout Asia, Africa, and Europe. Islam is especially important in Southwest Asia, North Africa, and some parts of Southeast Asia such as Indonesia and Malaysia. Islam is divided into two major sects, which are called Sunni and Shi'a (or Shiite). These sects disagree over the line of succession from the Prophet Muhammad. Sunni is the dominant sect in terms of number of followers, but in some countries, such as Iran, Iraq, and Syria, Shi'a is the dominant sect.

Islamism is an anti-colonial, anti-imperial, and anti-core movement that attempts to resist the globalizing and modernizing forces reaching Islamic countries largely from core countries. Islamists seek to establish a theocracy, in which secular government and religion are united in a common leadership. Islamists see globalizing influences and non-Islamic culture as threatening Islam as a religion as well as the well-being of Muslims. Though Islamism is a radical and militant movement, it is only one of many movements within Islam.

Culture and Identity

Other forms of cultural identity—in addition to language and religion—are also important. An emerging field within cultural geography is the study of sexuality. The Gender and Sexuality Issues website, linked to the textbook's website, contains a great deal of information relevant to the geographical analysis of these issues, including material on gender studies and gay studies. **Sexuality** is a set of practices and identities that a given culture considers related to each other and to those things it considers sexual acts and desires. Prostitution—or sex work—is an activity that can be analyzed geographically. Many cities have areas where prostitution takes place. Female sex workers servicing lower income clients tend to operate within public spaces, whereas sex workers servicing wealthier clients have very little public visibility or identity.

Another topic explored by geographers studying sexuality is the spatial constraints on homosexuality, and how gays and lesbians respond to and reshape these constraints. Gay and lesbian consumerism and political action are both responses to spatial and social constraints. Gay and lesbian consumerism involves the support of gay and gay-friendly businesses, often leading to identifiable gay spaces within some cities. Gay political activism is often concerned with what activities or values can be practiced or displayed in public spaces—it is often an attempt to de-heterosexualize space. Like other minority groups, many gays and lesbians resist the restricted uses of space forced upon them by the dominant social group. Through both consumerism and activism, gays and lesbians resist the control of space by culturally dominant groups.

Cultural identity can also be explored through ethnicity. **Ethnicity** is a socially created system of rules about who belongs and who does not belong to a particular group based upon actual or perceived commonality. Territory is usually the basis for ethnic group identity, even within present-day interpretations of ethnic identity, such as Hispanic and African-American. Hispanics share a common territorial heritage in Spain, while African-Americans share a common territorial heritage in Africa. Geographically, cultural or ethnic groups may be spatially separated—voluntarily or involuntarily—from other groups within a territory. And like the gay and lesbian case described above, ethnic groups may attempt to use or control space through consumption patterns and political activism. Ideas about race have also been used to understand the shaping of places and responses to these forces. **Race** is a problematic classification of human beings based on skin color and other physical characteristics. Race is a *social* construct—there is no biological basis for racial classification. In many respects, race is a social invention and an attempt to make ethnicity "natural" rather than something that is more properly understood as cultural.

Gender is another form of identity studied by geographers. Gender refers to the *social* differences between men and women rather than the *anatomical* differences that are related to sex. Cultural systems that identify men as more valuable than women can make it difficult for women to withstand natural disasters. Gender differences may also be mediated by social standing—women from wealthier households are less likely to experience shortages of food than are women from poorer households.

Globalization and Cultural Change

The focus of this chapter has been on how globalization changes culture. Globalization affects different groups in different ways, and these groups have likewise responded in different ways. Globalization has made many parts of the world similar to each other. Certain cultural products—especially aspects of U.S. popular culture such as jeans, clothing with college and sports logos, movies and television programs, fast foods, popular music, and many other things—are now found almost everywhere in the world. In this sense, many parts of the world resemble each other. These cultural products symbolize a lifestyle associated with luxury, youth, fitness, beauty, and freedom—the characteristics that advertising promotes and which many people desire. In some respects, this globalization of culture has brought people closer together—people all over the world now share many aspects of culture. Yet often these cultural objects are interpreted and used in different ways within different cultures.

Many cultures have resisted globalization to a greater or lesser degree. This resistance, as well as different interpretations and uses given to global culture, have ensured that cultural differences remain. The world is not yet dominated by a global culture. Culture is a central issue within the field of human geography. In order to understand the relationship between people and their environment, we must understand something about culture: what it is, how it is formed, and how it changes. Language and religion are two of the most important aspects of culture that help to shape cultural identity. These and other aspects of culture, such as gender, ethnicity, and sexuality, have been used as tools of resistance to globalizing and dominating cultural trends, helping to ensure that the world remains a place of great variety.

Review Questions

Some of the answers to these questions may be found only in the textbook, and not in the study guide.

1. The combination of traits characteristic of a particular group is a _____.

2. Judaism, Christianity, Islam, Buddhism, and Hinduism all emerged on the continent of _____.

3. _____ studies how space, place, and landscape shape culture at the same time that culture shapes space, place, and landscape.

4. The geographic origins or sources of innovations, ideas, or ideologies are known as:
    a) *genre de vie*
    b) rites of passage
    c) cultural hearths
    d) hinterlands

5. A characteristic and tangible outcome of the complex interactions between a human group and a natural environment is known as the _____.

6. An effort to protect regional and national cultures from the homogenizing impacts of globalization, especially the penetrating influences of U.S. culture, is known as:
    a) popular culture
    b) cultural nationalism

    c) folk culture

    d) cultural complex

7. _____ is the world's second largest religion, with about one billion adherents.

8. True or False: A cultural system is a single aspect of the complex of routine practices that constitute a particular cultural group.

9. _____ is a problematic classification of human beings based on skin color and other physical characteristics.

10. _____ is a shared set of meanings that are lived through the material and symbolical practices of everyday life.

11. Regional variations in standard languages are called:

    a) dialects

    b) pidgin languages

    c) lingua franca

    d) language branches

12. A _____ is a form of social identity created by groups who share a set of ideas about collective loyalty and political action.

13. A spatial dispersion of a previously homogeneous group:

    a) cultural hearth

    b) diaspora

    c) hajj

    d) cultural complex

14. _____ is a socially created system of rules about who belongs and who does not belong to a particular group based upon actual or perceived commonality.

15. Most languages in Europe and northern India belong to which language family?
_____

16. True or False: Popular culture refers to the traditional practices of small groups, especially rural people with a simple lifestyle.

17. A functionally organized way of life that is seen to be characteristic of culture groups, as described by the geographer Paul Vidal de la Blache, is known as _____.

18. A pilgrimage to Mecca made by Muslims is:
    a) jihad
    b) Ramadan
    c) hajj
    d) Sunni

19. The correct name for the nationalist movement sometimes known incorrectly as "Islamic Fundamentalism" is _____.

20. The two key components of a cultural system for most of the world's people are:
_____

21. Historical geography is usually associated with which British geographer? _____.

22. A means of communicating ideas or feelings by means of a conventionalized system of signs, gestures, marks, or articulate vocal sounds is called _____.

23. A _____ is a collection of individual languages that possess a definite common origin but have split into individual languages.

24. True or False: A language branch is a collection of individual languages believed to be related in their pre-historical origin.

25. A collection of individual languages that are part of a common branch, share a common origin, and have similar grammar and vocabulary is called a _____.

26. A person who practices Islam is called a _____.

27. A belief system and a set of practices that recognize the existence of a power higher than humans is a _____.

28. True or False: A cultural region is the area(s) within which a particular cultural system prevails.

29. The ceremonial acts, customs, practices, or procedures that recognize key transitions in human life such as birth, menstruation, and other markers of adulthood such as marriage, are:
    a) diasporas
    b) popular culture
    c) rites of passage
    d) dialects

30. True or False: Within Islam, "jihad" always refers to a violent struggle against enemies.

31. The spatial dispersion of a previously homogeneous group of people is referred to as a _____.

84   Chapter 5

32. A collection of interacting elements that taken together shape a group's cultural identity is a _____.

33. In Great Britain, the approach to understanding the human imprint on the landscape was known as _____.

34. _____ is the set of practices and identities that a given culture considers related to each other and to those things it considers sexual acts and desires.

35. The social differences between men and women, as opposed to their anatomical differences, are known as _____.

Conceptual Questions

1. Do you see yourself as belonging to a distinct cultural group (or groups)? What characteristics define that group?

_____
_____
_____
_____
_____
_____
_____
_____

2. What are some of the features of the natural and cultural landscapes in your community? How has the natural landscape shaped the cultural landscape, and how has the cultural landscape shaped the natural landscape?

3. How have religions and languages been used as instruments of resistance to globalization? Give specific examples.

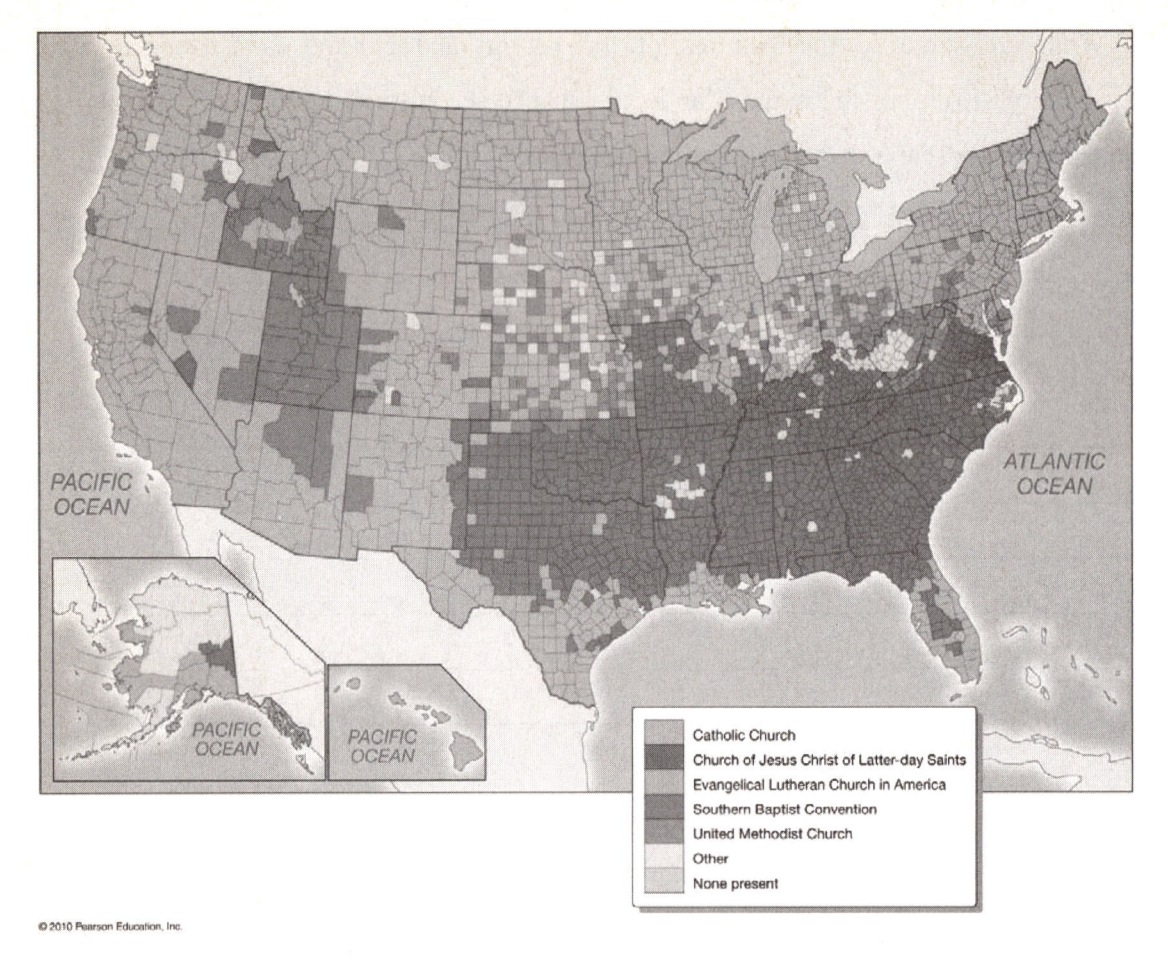

Figure 5.9

4. Using Figure 5.9, discuss how religion can be used to illustrate a cultural region.

5. How are measures of identity—such as sexuality, ethnicity, and gender—used as instruments of resistance to culturally dominant groups? How is space a part of this resistance?

_____
_____
_____
_____
_____
_____
_____
_____
_____
_____

6. Does your community show signs of globalization and the impact of global culture? What are these signs? Has your community (or parts of it) resisted globalization? If so, how?

_____
_____
_____
_____
_____
_____
_____
_____
_____
_____

7. What are the advantages and disadvantages of the globalization and standardization of culture? Would you prefer to live in a community in which everyone were the same (no

88 Chapter 5

differences in language, religion, sexuality, and so forth) or one in which such differences exist?

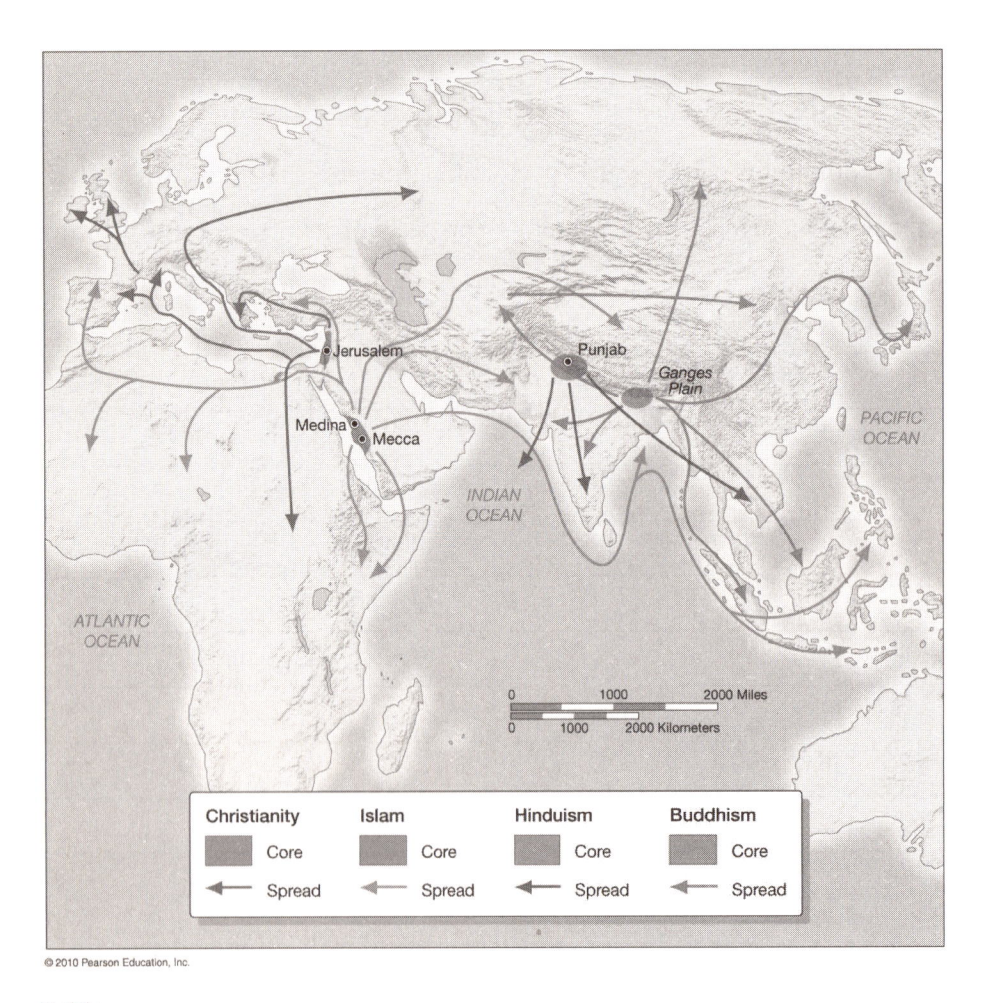

_____
_____
_____
_____
_____
_____
_____
_____
_____
_____

Figure 5.11

8. Figure 5.11 shows the origins and diffusion of four major religions. Discuss why these religions may have originated where they did, and why they diffused to the areas shown, rather than to other areas.

6

Interpreting Places and Landscapes

Chapter Objectives

The objectives of this chapter are to illustrate:

1. How different groups of people experience landscape and place differently
2. How ordinary landscapes differ from symbolic landscapes
3. How all landscapes can be understood

Chapter Notes

Behavior, Knowledge, and Human Environments

In addition to understanding how the environment shapes and is shaped by people, geographers also seek to identify how it is *perceived* and *understood* by people. The knowledge that people acquire about their environment helps shape their attitudes and behaviors. People acquire this knowledge through experience, which is in turn filtered by personal and group characteristics such as gender, age, religious beliefs, where we live, and other factors. Knowledge of our environment—and the ways it shapes us—changes as we get older and have more experiences. For example, children will relate to their environment in a way different from the way that adults relate to it. Children acquire knowledge of their environments in various ways, including playing, doing chores, and from information given to them by adults. When children get older and attend school, a new factor—formal education—is added, and helps shape knowledge of the environment.

Place-Making

Some social scientists believe that human beings, like other animals, have an innate sense of territoriality. **Territoriality** is the persistent attachment of individuals or peoples to a specific location or territory. The study of territoriality is part of the larger field of **ethology**, which refers to the formation and evolution of human customs and beliefs and to the behavior of an animal in its environment. Ethologists argue that people have territorial impulses or traits that govern their behavior. We can see examples of territorial behavior on the individual level by noticing how people make territorial claims to public spaces such as "their" seat on a bus or in a classroom.

Groups may also make territorial claims to selected places. The study of the social and cultural meanings that people give to personal space is known as **proxemics**. People make personal claims to public spaces such as seats on buses or classrooms, as noted above. In these examples, regular use, or always sitting in the same seat, gives an individual a claim to the seat. Personal spaces can also be marked, for example by placing a backpack or jacket on the seat to indicate that it has already been taken. People also have bubbles or areas of personal space around them. Most people have a sense of when another person is too close to them, and this sense of closeness may vary between cultures.

Territoriality helps serve three social and cultural needs: the regulation of social interaction; the regulation of access to people and resources; and the provision of a focus and symbol of group membership and identity. Territoriality serves these ends because it helps people classify others, for example, as persons that may or may not enter their territory. It is much easier to classify people in this spatial way than it is to classify by personal or social criteria.

People can establish bonds with each other by sharing a sense of place. A **sense of place** is the feeling evoked among people as a result of the experiences and memories that they associate with a place, and to the symbolism that they attach to it. We can identify people as "locals" or as "outsiders." Common bonds may exist between people who consider themselves "Southerners" or "Texans," bonds that are not formed with people who do not share this same sense of place. The sense of place may be reflected in shared dress codes, speech patterns, public behavior, and so on. Texans, for example, may recognize each other by their accent, indicating who is a Texan and who is not; that is, who shares the sense of Texas space and who does not.

How do people derive meaning from space? People derive a large part of meaning from experiences, which are in turn filtered by the five senses, by the brain and individual personality, and by culture. Cognitive images are the way we see places in our imagination—how we visualize or remember them. Cognitive images can both simplify and distort the real world. People tend to organize, or make sense of, their spatial world by dividing it into a number of categories, including paths, edges, districts, nodes, and landmarks. Cognitive images may be distorted, generally by a lack of information. Once we leave our immediate surroundings, we generally do not know the exact details of places. Our images of places may be influenced by fragmentary and often biased information from other people, books, magazines, television, films, and other sources. For example, the Hudson River School of 19th-century American artists created a particular image of the American landscape—see the link to information about the Hudson River School in the textbook's website. Our images of places also reflect our own personalities, values, and tastes.

We learn about our environments through experience, and the cognitive images that we develop from these experiences shape our behavior. These cognitive images are constantly changing as our experiences multiply. The more knowledge and the more experiences of a place we have, the more detailed our cognitive images will be. Cognitive images shape behavior in a variety of ways. For example, research on shopping behavior indicates that people do not always shop at the nearest or cheapest store. Often they will shop instead at places for which they have a positive cognitive image, often reflected in the convenience of the place in such amenities as parking or easy access, or because the shopping environment is seen as more attractive. Therefore, the developers of shopping malls and other shopping districts pay close attention to the cognitive images that people have of the place, and will attempt to create the atmosphere expected by their clientele.

Cognitive images also influence migration patterns. Within the United States, people identify certain regions as more or less desirable to live in based on cognitive criteria. Often this assessment of desirability is linked to **topophilia**, the emotions and meanings associated with particular places that have become significant to individuals. Many people have a strong attachment to their home region, and would dislike leaving it even if better economic opportunities opened up elsewhere.

Places and regions throughout the world are increasingly seeking to influence the ways that they are perceived by tourists, businesses, media firms, and consumers (see the link to the Politics of Travel site in the textbook's website). Sometimes these places will consciously re-interpret, re-imagine, design, package, and market themselves. This process often involves manipulating the visual imagery of the place in order to appeal to selected groups. For example, the city of Santa Fe, New Mexico, was consciously altered in the 1920s from a city with many Victorian and "American" buildings to a city that its promoters thought more closely approximated the architecture of the Spanish colonial and Pueblo Indian cultures. Artists, intellectuals, and other residents who had a stake in transforming the city deliberately planned this alteration of the city's image and style. Santa Fe's Spanish-Pueblo image has made it a distinctive place that is popular with tourists.

Landscape as a Human System

Landscape is the product of human interaction with the environment. Ordinary landscapes (or vernacular landscapes) are the everyday landscapes that people create in the course of their lives. Parking lots, suburban housing developments, and farms are all examples of ordinary landscapes. Symbolic landscapes are representations of particular values and/or aspirations that the builders and financiers of those landscapes want to impart to a larger public. For example, many government buildings in the United States, such as court houses, are built in a neoclassical style, which symbolizes not only power and authority, but also the Greek and Roman classical ideals of democracy and republicanism. Often banks are built in a similar style, because this style—due to its government associations—also symbolizes stability and security. **Derelict landscapes** are those that have experienced abandonment, misuse, disinvestment, or vandalism. Inner city slums are often derelict landscapes, as are abandoned farms.

Different people may perceive the same landscape in different ways. The **humanistic approach** in geography is concerned with individual perception of the landscape, and places the individual person at the center of analysis. The humanistic approach attempts to understand how people perceive their worlds, often in a subjective, psychological way. Another approach to the study of landscape is to consider landscape as text. **Landscape as text** is the idea that landscapes contain meanings that can be read and written by groups and individuals. In the same way that we read books, we can read landscapes. And, as in the case with books, sometimes different people will draw different conclusions or meanings from a given landscape. Landscapes are created or "written" by some people, while others interpret or "read" them. Numerous websites document and explore the world's landscapes. The textbook's website has links to several, including those on the Great Lakes, the Arctic, the Lower Mississippi Valley, and New York City.

Coded Spaces

Landscapes can be interpreted as texts in the same way that one reads and interprets a book. As with a book, we read landscapes by making sense of symbols and signs embedded in them. This practice of writing and reading signs is known as **semiotics**, the science of signs. Many signs are displayed in landscapes and these signs may have different meanings for those people who produce them and those who read them. For Americans, shopping is the second most important leisure activity after watching television. For many people, shopping is a kind of tourism rather than simply the fulfilling of basic material needs. Shopping malls, the most important type of retail space in the United States, are full of signs or semiotic systems. These signs often indicate who is welcome at the mall, and who is not.

In many communities, shopping malls have become the centers of public activity, replacing Main Street as the core of urban (or suburban) life. But malls differ from Main Street in one important respect: they are privately owned, and the owners may not only control the writing of signs embedded in the malls, they may also control access, designate what is acceptable behavior, keep people under surveillance, and remove people deemed undesirable from their property. The primary purpose of malls, unlike Main Street, is consumption – convincing people to buy products and services. By using signs and symbols associating consumption with the Good Life, the producers of mall semiotics can achieve their ends.

Religious places can also be read and decoded. **Sacred spaces** are areas recognized by individuals and groups as worthy of special attention as a site of special religious experiences or events. Sacred spaces may be the sites of pilgrimages, or journeys, undertaken by people to demonstrate devotion or to renew their faith—for more information follow the link in the textbook's website to the Sacred Landscapes site. Sacred spaces, like other cultural spaces, have special meanings to believers, and may have different meanings or no meanings to non-believers.

Place and Space in Modern Society

Since the 1980s, many commentators have noted a broad shift in cultural activities and values. This shift is sometimes characterized as a shift from Modernity to Postmodernity. **Modernity** is a forward-looking view of the world that emphasizes reason, scientific rationality, creativity, novelty, and progress. This worldview, or way of seeing the world, originated with the European Renaissance and was consolidated into a distinct paradigm in the Age of Enlightenment (the 18th century). Science, citizen politics, and a capitalist economy characterize Modernity, as do distinctive manifestations of these ideas in literature, art, and architecture.

After World War II, global society began to change in significant ways, largely due to the impact of new technologies, communications systems, and media. These changes are so distinctive that many cultural analysts believe that we have entered a new paradigm or worldview, in the same way that the Renaissance was different from the Middle Ages. This new worldview is called **Postmodernity**, and it emphasizes openness to a range of perspectives in social inquiry, artistic expression, and political empowerment. Many Postmodernists challenge the dominant position of rational thinking, science, progress, and capitalism in today's society. Postmodernity also focuses on living for the moment; it is a "society of the spectacle" in which image and appearance, and the symbolic meanings of things, become more important than production and use of material objects. Style, rather than function, is Postmodernity's emphasis. Nostalgia, or the desire for things and ways of the past, is also a feature of Postmodernity—consider the evocations of small town America in such artificial, image-based creations as Disneyland and pioneer villages, and in "retro" fashions.

Globalization has brought new styles and images to all parts of the world. This diffusion of styles and images is accompanied by the increased importance of consumption, not only of material goods, but also of images and appearances. Jeans are no longer just jeans—workingmen's trousers—but statements of different kinds of styles, small semiological systems, in which the wearing of Levi's, Wrangler, or Guess jeans sends different messages and contains different meanings. Consumers are not just buying a pair of pants, but an *image* connected with them. This is the essence of Postmodernity. Postmodernists are characterized by **cosmopolitanism**, an intellectual and aesthetic openness toward divergent experiences, images, and products from different cultures. Cosmopolitans can be at home in any culture. Landscapes are symbolic places, full of coded meanings that can be read and interpreted. Likewise, landscapes can be written, and meanings can be encoded into them. Who is able to encode meaning, and how people decode these meanings, reflect power relationships within culture. The transition from Modernity to Postmodernity will continue as new meanings in landscapes are constantly being written and read.

Review Questions

Some of the answers to these questions may be found only in the textbook, and not in the study guide.

1. _____ is an intellectual and aesthetic openness toward divergent experiences, images, and products from different cultures.

2. According to the textbook, Postmodernity is characterized by what three emphases?

3. Landscapes that have experienced abandonment, misuse, disinvestment, or vandalism are called:

    a) ordinary landscapes

    b) symbolic landscapes

    c) derelict landscapes

    d) vernacular landscapes

4. Throughout most of the 20th century, _____ was the philosophy or view that was most influential in shaping the interdependencies among culture, society, space, place, and landscape.

5. The _____ places the individual—especially individual values, meaning systems, intentions, and conscious acts—at the center of analysis.

6. Instead of focusing on economic and scientific progress, Postmodernity is largely _____-oriented.

7. True or False: The idea that landscapes can be read and written by groups and individuals is called Postmodernity.

8. The architect of the Brazilian capital city of Brasilia, Lucio Costa, used the symbol of a _____ to indicate the sacred nature of the new capital.

9. The forward-looking view of the world that emphasizes reason, scientific rationality, creativity, novelty, and progress is called:

    a) Modernity

    b) Postmodernity

c) ethology

d) topophilia

10. True or False: Postmodernity: a view of the world that emphasizes an openness to a range of perspectives in social inquiry, artistic expression, and political empowerment.

11. The study of the social and cultural meanings that people give to personal space is known as:
    a) ethology
    b) proxemics
    c) landscape as text
    d) cosmopolitanism

12. An area recognized by individuals or groups as worthy of special attention as a site of special religious experiences or events is called _____.

13. Among the kinds of cognitive images, _____ are those that are physical reference points such as buildings or monuments.

14. The practice of writing and reading signs is:
    a) semiotics
    b) ethology
    c) topophilia
    d) humanistic approach

15. _____ is the persistent attachment of individuals or peoples to a specific location or territory.

16. "Main Street, U.S.A.," is an example of a _____ landscape.

17. True or False: Topophilia is the complex of emotions and meanings associated with particular places that, for one reason or another, have become significant to individuals.

18. The humanistic approach to geography places the _____ at the center of analysis.

19. The landscape as text conceptualization suggests that, rather than being ready-made, landscapes can be _____ and _____ by individuals and groups.

20. Territoriality provides a means of accomplishing what three cultural and social needs?

_____
_____
_____

21. The scientific study of the formation and evolution of human customs and beliefs is _____.

22. _____ is the practice of reading or writing signs.

Conceptual Questions

1. Give examples of ordinary, symbolic, and derelict landscapes in your community or region. What makes these landscapes ordinary, symbolic, or derelict?

_____
_____
_____
_____

2. Consider the shopping malls or districts in your community. How would you read the signs in this mall? What kind of image did the writers of the signs attempt to create?

3. In what places in the world would you like to live? What is appealing about these places? How does this appeal relate to your personality and cultural background?

4. What examples of human territoriality can you find in your community? How do these examples demonstrate territoriality?

5. What tourist attractions does your community possess? What makes them of interest to tourists? Does a semiological approach help to understand what makes these places of touristic importance?

6. Are there any sacred spaces in or near your community? How do you know they are sacred spaces? How are these spaces distinctive from ordinary spaces?

_____
_____
_____
_____
_____
_____
_____
_____

7. What is meant by Postmodernity? Is there any evidence of Postmodernity in your community?

_____
_____
_____
_____
_____
_____
_____
_____

8. Figure 6.8 shows a preference map of the United States. The map is based on the preferences for cities in which to live as expressed by students at a university in Virginia. Discuss why Virginia students might have these preferences. What would your own preferences be, and why?

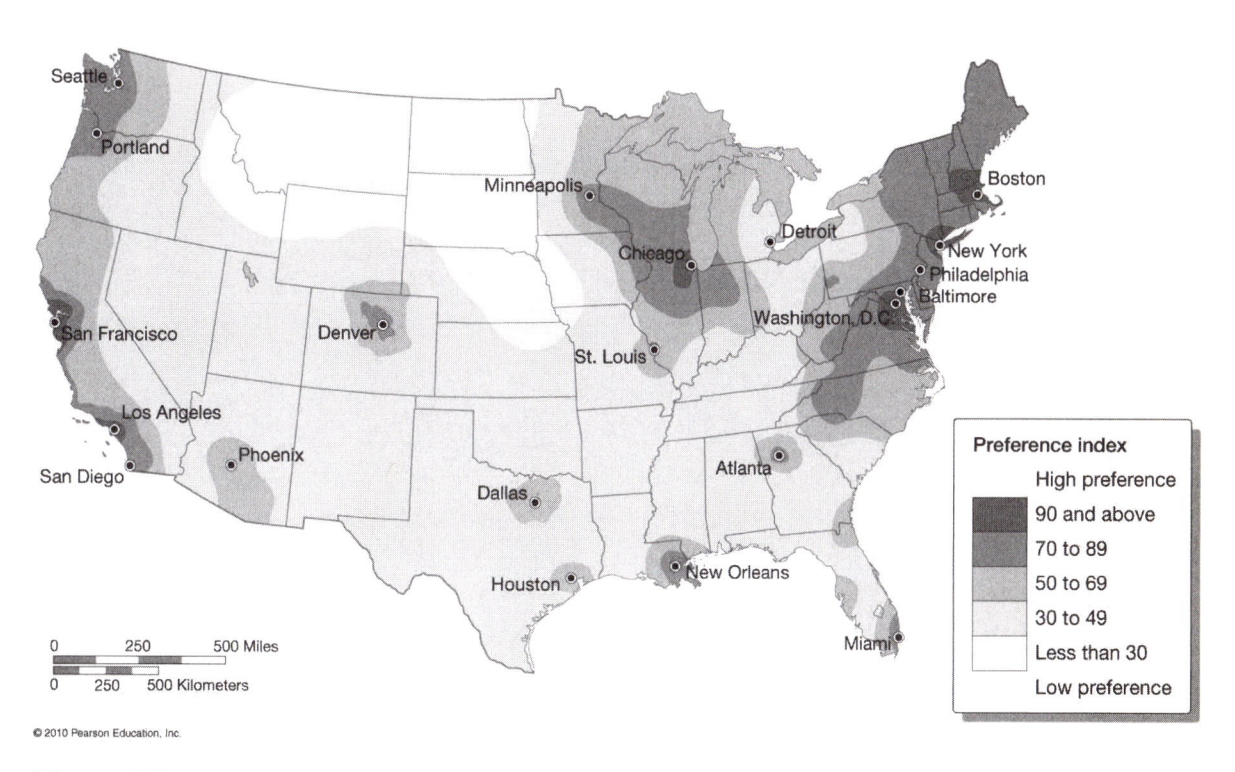

Figure 6.8

7

The Geography of Economic Development

Chapter Objectives

The objectives of this chapter are to:

1. Examine the unevenness of economic development in various parts of the world
2. Investigate the economic structure of countries and regions, and explore the various stages of economic development
3. Survey principles of commercial and industrial location and how they affect economic interdependence
4. Examine core-periphery patterns and how they are created
5. Explore the pleasure periphery

Chapter Notes

Patterns of Economic Development

Economic development is often discussed in terms of levels and rates of change in prosperity, as reflected in statistics measuring productivity, incomes, purchasing power, and consumption. But prosperity is only one element in the larger concept of economic development. Processes of change involving the nature and composition of the economy of a particular region as well as increases in the overall prosperity of a region are also important aspects of economic development, as are improvements in the basic conditions of life. Processes of change in a region's economy can take three forms: changes in the structure of the region's economy (for example, a shift from agriculture to manufacturing); changes in forms of economic organization within the region (for example, a shift from socialism to free-market capitalism); and changes in the availability and use of technology within the region. Each of these three types of change can affect the overall prosperity and conditions of life in a region.

Geographically, economic development is uneven. Different regions have different levels of economic development. The core regions—North America, Europe, and Japan—are the most economically diversified, have the highest levels of productivity, the most advanced technologies, and the highest levels of prosperity. The core regions are sometimes known as the *developed* regions, and, formerly, as the First World. The peripheral and semiperipheral regions of the world are sometimes known as the *developing* or *less-developed* regions, and, formerly, as the Third World (the Second World referred to communist countries).

Levels of economic development are usually measured by economic indicators such as gross domestic product and gross national product. **Gross domestic product (GDP)** is an estimate of the total value of all materials, foodstuffs, goods, and services that are produced by a country in a particular year. In order to make comparison easier, this gross domestic product is sometimes divided by the number of people in the country to give gross domestic product per capita. **Gross national product (GNP)** is similar to gross domestic product, but also includes the value of income from abroad; for example, foreign earnings of domestic companies. These indicators are used to distinguish core countries from peripheral and semiperipheral ones. Two websites, the Bureau of Economic Analysis and the World Bank's World Tables Dataset Guide, which are linked through the textbook's website, provide a good source of economic data, including GDP and GNP, for all countries.

The unevenness of economic development is also reflected in gender equality. According to a gender-sensitive index developed by the United Nations, in no countries are women better off than men. This index measures the employment and wages of women compared to men. The general pattern that emerges from this index is that in core countries women are only slightly less well off than men, while in countries of the periphery women are far worse off. These figures reflect only women in the salaried labor force, and not the work performed by women in the home or in subsistence agriculture. In addition to differences in levels of economic development *among* countries, one can observe differences *within* countries as well.

Resources and technology both influence patterns of economic development. The presence of key resources—cultivable land, energy sources, and valuable minerals—can influence the economic development of a country or region. Natural resources are unevenly distributed throughout the world. Some countries, and some regions, have more cultivable land, more energy sources, and more valuable minerals than do others. While the presence of these resources does not guarantee prosperity, it can be an important contributing factor. Countries may also offset a lack of resources through trade—Japan is an excellent example of this. Levels of technology are also closely tied to the presence of natural resources. The substitution of some resources for others—such as oil for coal, or synthetic rubber for natural rubber—can change the importance of a region's natural resource endowment. Regions that are dependent on a single resource are especially vulnerable to changing world conditions. The use of other metals, such as aluminum, in place of tin, for example, has negatively affected the economies of countries such as Bolivia that heavily depended on tin exports.

Economies have different components, or kinds of economic activity. The relative share of these kinds of activity determines the *economic structure* of a country or region. **Primary activities** are those concerned directly with natural resources of any kind, such as mining, agriculture, or forestry. **Secondary activities** are those that process, transform, fabricate, or assemble the raw materials derived from primary activities, and include textile manufacture, steelmaking, and automobile assembly. **Tertiary activities** are those involving the sale and exchange of goods and services, such as retail stores, accounting, and entertainment. **Quaternary activities** are those dealing with the handling and processing of knowledge and information, including data processing, education, and research and development.

Primary activities are dominant in most of the peripheral countries, while secondary activities are more important in core and semiperiphery countries. Tertiary and quaternary activities are almost entirely restricted to the wealthiest countries of the core. Figure 7.10 in the textbook illustrates the importance of primary sector activities in different countries. Most of the countries where the primary sector dominates are in the poorest countries of the periphery. Immanuel Wallerstein's world-systems model, discussed on the Core-Periphery website (linked through the textbook's website) describes the origins of cores and peripheries and the relations between them. The variations in economic structure noted above reflect *geographical divisions of labor*. The geographical division of labor reflects who does what: some countries are producers of raw materials, some are manufacturers, and others are *postindustrial*, specializing in services and information activities. The highest levels of economic development, as measured by GDP, are associated with the last group.

The geography of international trade is very complex. International trade is based on a few **trading blocs**, which are groups of countries with formalized systems of trading agreements. There are four main trading blocs: Western Europe and its former colonies; North America and some Latin American states; the countries of the former Soviet Union and its allies; and Japan along with East Asia and some oil-exporting countries. Some countries are not heavily involved in the global system of trade, a condition known as **autarky**. In many peripheral countries, much or all export earnings are used to pay the interest and principal on international debt. Part of this situation is due to the **international division of labor**, in which some countries specialize in particular products for export. The relationship between price and the demand for a product is known as **elasticity of demand**. If elasticity is high, then a small change in price will have a large change in the demand for the product.

Some peripheral countries have attempted to change their position in the international division of labor by attempting to manufacture their own products, rather than importing them from core countries, a policy known as **import substitution**. Some scholars have interpreted economic development as a process involving distinct stages. W.W. Rostow, an American economist and political adviser, developed a model based on *stages of growth*, arguing that every country or region passes through five stages, from traditional society to high mass consumption. Rostow's model is depicted in Figure 7.16 of the textbook. Though highly influential, this Modernist model of development is rather simplistic and based on certain assumptions concerning factors that are not always present. Rostow's model argues for the primacy of *internal* factors in a country's development, and largely ignores the impact of external forces and globalization. The model suggests that more investment capital and free trade are the solutions to development problems in peripheral countries.

Pathways to Regional Development

Patterns of economic development result from the principles of location and from historical circumstances. In a given location, the way of organizing economic activity in the past may shape the way economic activity is organized in the present. This historical connection between past and present is known as **geographical path dependence**. Economic history is important: having the **initial advantage**, or having an early start, can positively affect a firm's or industry's prosperity. Thus, for example, the Santa Clara Valley of California—sometimes known today as Silicon Valley—had an initial advantage in high-tech electronics, something that is difficult for other regions to achieve in the same industry.

Regional cores of economic development are created cumulatively, following some initial advantage. The local advantage is reinforced by the principles of agglomeration and localization discussed above, and this gradual buildup of advantages is known as **cumulative causation**. The high-tech electronics industry in Silicon Valley had an initial advantage in that it was located near a major university—Stanford—where many of the founders of electronics firms studied or worked. As these firms—such as Hewlett-Packard—grew in size, others started to develop nearby, creating forward and backward linkages. The high educational level of the population, the genial climate, the availability of start-up capital, and the related research activities of the nearby universities drew firms to the area. Today many other high-tech firms, such as biotechnology industries, are located in the area, taking advantage of the same conditions that the electronics firms enjoy. Regions may compete with each other in the same industry, and the relative growth of an industry in one region may lead to the decline of that industry in another region. These negative impacts are known as **backwash effects**. In the United States, the availability of cheap labor and power in the South led to the growth of a textile industry there, and the corresponding decline of the textile industry in New England.

Core-periphery economic relations can be altered in a variety of ways. Sometimes the relative growth of one region may have a positive impact on other regions, through a trickle-down effect. These trickle-down effects are usually known as **spread effects**. The economic growth of South Korea, for example, is partly attributable to the economic growth of its neighbor Japan. Agglomeration can also have negative effects, or **agglomeration diseconomies**. These result from urbanization and the concentration of industry, and may be reflected in higher costs for land and labor, transportation congestion, and higher taxes needed to support increased social services.

Changes in technology and competition between countries may result in the relocation of industries. **Deindustrialization** is a relative decline in industrial employment in core regions, usually attributed to the growth of the same industry in peripheral regions. Thus, many manufacturing firms in the United States have relocated from the northern Rustbelt to the southern Sunbelt. The taking apart or dismantling of old core industrial regions may result in new investment capital being available for new areas. This is known as **creative destruction**, in which investment is withdrawn from one area and reinvested in another. Core-periphery economic relations can also be altered by government intervention. The best-known example of this process is the growth of Japan after World War II, which was aided by government assistance. An important government response to economic development is the creation of **growth poles**, or economic activities that are centered around one high-growth industry. Biotechnology, for example, is a high-growth or "propulsive" industry today, and many local and regional governments have attempted to achieve economic development centered around this industry.

Globalization and Economic Development

The dynamics of economic globalization rest on flows of capital, knowledge, goods, and services among countries. In 2006, private companies invested more that $1.3 trillion in businesses outside of their own countries. This **foreign direct investment** is twenty times more than what it was thirty years earlier. Almost half of this investment was targeted for peripheral and semiperipheral countries. The world economy has become increasingly globalized and integrated. Part of this process has been the development of **transnational corporations**, which conduct their business in many countries. Many of the world's largest corporations are **conglomerate corporations**, which means that they are diversified into a variety of economic activities and industries. Some of these transnational conglomerate corporations are economically larger than countries such as New Zealand or Norway.

Transnational corporations have created a *global assembly line*, in which different aspects of their activities are carried out in different countries. For example, labor-intensive work can be done in peripheral countries where labor is cheap, while final assembly and finishing touches can be done near the markets for the products, usually in core countries. Many automobile firms produce and assemble different components of their product in different locations—the components of the Ford Escort, for example, are made and assembled in 15 countries across three continents. The automobile and garment industries make extensive use of a global assembly line, as Figure 7.20 in the textbook makes clear. Corporations may also form strategic alliances with other corporations in similar or related fields. These alliances allow corporations to reach new markets, to obtain new information and technology, to reduce the costs of product development, and to spread the costs of market research. Transnational corporations seek to take advantage of differences—for example, in labor costs—between countries. Nike, the American footwear and clothing manufacturer, used to produce its products in factories located in the United States and United Kingdom. As labor costs in these countries increased, Nike subcontracted its manufacturing to suppliers in lower-cost countries such as South Korea and Taiwan. Then, as labor costs rose in these countries, Nike switched to even lower-cost countries such as China, Indonesia, and Thailand.

Transnational corporations have been an important element in the transition from Fordism to Neo-Fordism. **Fordism**, named after the American automobile manufacturer Henry Ford, refers to mass production based on the assembly line and on mass consumption based on higher wages and advertising techniques. In **Neo-Fordism**, the basis of Fordism has been modified by more flexible systems of production and consumption. Flexible production systems involve greater flexibility both within firms and between them. Some countries have encouraged transnational corporations to locate manufacturing facilities in their countries by providing tax and other incentives. For example, Mexico has legislation allowing foreign (mainly American) companies to locate within a few miles of the United States border—in centers known as *maquiladoras*. American companies can take advantage of lower labor costs in Mexico while being assured that they can import their products duty-free into the United States. Some countries have established **export-processing zones (EPZs)**, which grant foreign corporations incentives such as exemption from duties and tariffs, as well as subsidized rents and taxes, in return for the employment that the foreign corporations provide.

Banking, finance, and business services have also become global industries. One of the consequences of this globalization is that the majority of international transactions have nothing to do with the trade in goods and services. Instead, banking and finance have become ends in themselves, with profits derived from speculation in global financial markets. The increased size of banks and investment firms participating in these activities means that they are now able to control many aspects of economic development in peripheral countries and regions. (Yet, despite the dominance of giant banks, small microcredit institutions, which make loans to the world's poorest people, are also flourishing—see the Enterprise Development and Virtual Library on Microcredit websites linked through the textbook's website.)

Despite new communications technologies, economic activity and growth has become increasingly localized, or concentrated in specific places. Nevertheless, there has been some dispersion of service activity to peripheral areas, mainly in the form of "back-office" operations. Back-office operations include record keeping and other activities that do not involve much contact with clients or customers, and may be theoretically located anywhere. These back-office operations are normally located in places with low labor costs. One example is customer service operations in the United States making use of toll-free 800 numbers. When dialing an 800 number, one never knows what geographical location one is calling. Often the person answering the telephone may be working in a region where labor costs are low.

Most financial services are concentrated in a few large cities, where specialized infrastructure is available and where bankers and financiers may meet each other personally. These large financial centers—New York, London, Tokyo, Paris, and Frankfurt—are **world cities**, in which a disproportionate part of the world's most important business—economic, political, and cultural—is conducted. Some financial needs, such as secrecy and shelter from taxation and regulation, are not met by the metropolitan concentration of banking and finance activities. Such needs can often be met by **offshore financial centers**, usually small islands and countries with limited or no taxation and often loose regulations governing deposits and other transactions. Luxembourg, Switzerland, and the Cayman Islands are examples of offshore financial centers.

Tourism is an economic activity that has flourished in parts of the world lacking other economic activities. Tourism is also important in core regions that do possess economic links to the world system. Today tourism is one of the world's largest industries, involving people in core, semiperiphery, and periphery countries. Tourism can have both positive and negative economic effects: it can provide income and revenue as well as employment, but it can also lead to economic dependency on the tourism industry, which is often unstable because tourists' tastes change rapidly. Tourism can also have both positive and negative social and environmental effects. Tourism can help sustain indigenous lifestyles, regional cultures, and arts and crafts, and provide incentives for the protection of wildlife and historic buildings.

Tourism can also alter and destroy indigenous cultures and lead to environmental problems and pollution. Plenty of examples of both kinds of effects can be found in the United States alone. The demand for wilderness areas and parks has led to the establishment of national parks and monuments, yet some of these parks are now so overrun with people that the environment has been severely stressed. Tourists have injected money into Native American economies, while at the same time negatively affecting the privacy and culture of these peoples. In sum, tourism is a mixed blessing.

Economic development may be achieved through various pathways, as we have seen in this chapter. Some countries and regions have developed through a traditional stages-of-growth process, while others have prospered through establishing themselves as centers for back-office activities, as offshore financial centers, or as venues for alternative tourism. Thus, there are many roads to economic development. Even though the world has become more interdependent, economic development is uneven—core-periphery contrasts continue to exist, as do contrasts between rich and poor within and between countries.

Review Questions

Some of the answers to these questions may be found only in the textbook, and not in the study guide.

1. Cost advantages that accrue to individual firms because of their location among functionally related activities are called:
    a) agglomeration effects
    b) agglomeration diseconomies
    c) ancillary activities
    d) backwash effects

2. _____ are activities such as maintenance, repair, security, and haulage services that serve a variety of industries.

3. The gap between the world's rich and poor is getting _____.

4. True or False: Autarky refers to states that do not contribute significantly to the flows of imports and exports that constitute the geography of trade.

5. _____ is a vision of development that seeks a balance among considerations of economic growth, environmental impacts, and social equity.

6. The negative impacts on a region (or regions) of the economic growth of some other region are called:

    a) ancillary activities

    b) creative destruction

    c) backwash effects

    d) neo-Fordism

7. The maximum population that can be maintained in a place with rates of resource use and waste production that are sustainable in the long term is _____.

8. True or False: Monetarism is the withdrawal of investments from activities (and regions) that yield low rates of profit in order to reinvest in new activities (and new places).

9. The situation in which countries have to constantly borrow in order to finance development is known as:

    a) monetarism

    b) autarky

    c) inflation

    d) debt trap

10. _____ is a measure of the human pressures on the natural environment from the consumption of renewable resources and the production of pollution.

11. True or False: Quaternary activities are economic activities that are concerned directly with natural resources of any kind.

12. The degree to which levels of demand for a product or service change in response to changes in price is called _____.

13. The mass production system known as Fordism was named after _____.

14. Altria and Nestlé, because of their diversified activities, are examples of _____ corporations.

15. What are the three main indicators of economic development?
_____
_____
_____

16. True or False: Deindustrialization is a relative decline in industrial employment in core regions.

17. Small areas within which especially favorable investment and trading conditions are created by governments in order to attract export-oriented industries are called:
    a) export processing zones (EPZs)
    b) localization economies
    c) newly-industrializing countries
    d) trading blocs

18. _____ are economic activities that process, transform, fabricate, or assemble the raw materials derived from primary activities, or that reassemble, refinish, or package manufactured goods.

19. _____ are cost savings that result from circumstances beyond a firm's own organization and methods of production.

20. Cities in which a disproportionate part of the world's most important business—economic, political, and cultural—is conducted are called _____.

116   Chapter 7

21. _____ is the total of overseas business investments made by private companies.

22. _____ is the industry that is the world's largest non-agricultural employer.

23. _____ is an estimate of the total value of all materials, foodstuffs, goods, and services that are produced by a country in a particular year. _____ is similar, but in addition includes the value of income from abroad.

24. True or False: Growth poles are economic activities that are deliberately organized around one or more high-growth industries.

25. The specialization by countries in particular products for export is known as:
   a) elasticity of demand
   b) international division of labor
   c) import substitution
   d) cumulative causation

26. The doctrine of macroeconomic management that regards the money supply as the most important determinant of economic stability is:
   a) autarky
   b) dependency
   c) Fordism
   d) monetarism

27. The negative economic effects of urbanization and the local concentration of industry are known as _____.

28. Give four examples of newly industrializing countries.

_____
_____

29. OECD is the acronym for:
_____

30. The application of supply-side economics to the management of the U.S. economy in the 1980s was sometimes known as _____.

31. _____ are determined by the ratio of the prices at which exports and imports are exchanged.

32. What are the four major trading blocs in the world today?
_____
_____

33. The underlying framework of services and amenities need to facilitate productive activity is known as _____.

34. Companies that have diversified into a variety of economic activities, usually through a process of mergers and acquisitions, are called _____.

35. _____ is the critical importance of an early start in economic development; a special case of external economies.

36. True or False: Localization economies: cost savings that accrue to particular industries as a result of clustering together at a specific location.

37. Give four examples of offshore financial centers.
_____
_____

38. _____ are economic activities that deal with the handling and processing of knowledge and information.

39. The positive impacts on a region (or regions) of the economic growth of some other region are known as:
   a) spread effects
   b) terms of trade
   c) initial advantage
   d) agglomeration effects

40. Economic activities involving the sale and exchange of goods and services are called:
   a) primary activities
   b) secondary activities
   c) tertiary activities
   d) quaternary activities

41. _____ are companies that participate not only in international trade, but also in production, manufacturing, and/or sales operations in several countries.

42. The historical relationship between the present activities associated with a place and the past experiences of that place is known as _____.

43. _____ are Mexican "sister factories," located just across the border from the United States, that help foreign corporations take advantage of lower labor costs in Mexico.

Conceptual Questions

1. Economic development within countries and regions can be uneven. Can you give some examples of this unevenness within your own state or region?

2. Why are some countries more developed than others? Why, for example, is Switzerland a prosperous core country, while Bolivia is a poor peripheral country?

3. Collect data and draw a map of the world's offshore financial centers. What do they appear to have in common? Why are they located where they are?

4. What is the relationship between the status of women and economic development?

5. Explain the "stages of growth" model of economic development. What are some problems with this model?

6. What are the principal economic activities in your community? Are they primary, secondary, tertiary, or quaternary activities? Why did these activities develop in your community?

7. Is the economy of your community linked to the global assembly line? If so, how?

8. What are offshore financial centers, and how can they contribute to economic development?

9. What are some of the positive and negative features of tourism as an economic development strategy? Can you find evidence of these features in your own community?

8

Agriculture and Food Production

Chapter Objectives

The objectives of this chapter are to:

1. Understand traditional agricultural geography
2. Examine the agricultural revolution and its industrialization
3. Investigate the forces of agricultural globalization
4. Explore the social and technological change in global agricultural restructuring
5. Examine the relationship between the environment and agricultural industrialization

Chapter Notes

Traditional Agricultural Geography

Geography's approach to the study of agriculture is to understand the physical and human systems as interactively linked. Agricultural geographers have mapped the different factors—such as soil, temperature, and terrain—that shape agriculture, and they have mapped the distribution of different types of agriculture and the relationships between agriculture and other practices. Agricultural is a changing, dynamic activity. In the past four decades, the number of people employed in farming has declined, yet agricultural output has increased thanks to improvements in technology. Agriculture has been increasingly integrated into the global economic system, and it has become more directly linked to other sectors such as manufacturing and finance. Agricultural geographers also study the lifestyle and culture of different agricultural communities. Social scientists use the term **agrarian** to refer to those lifestyles and cultures that are linked to agriculture. Agrarian also refers to the type of tenure system, which determines who has access to land and what kind of agricultural practices will be used.

Agriculture itself is a science, an art, and a business directed at the cultivation of crops and the raising of livestock for sustenance and for profit. Agriculture is a good illustration of the interactions between people and the natural landscape, each affecting the other. Before agriculture was first invented or discovered by human beings, **hunting and gathering** was the mode of subsistence. People supported themselves by hunting and fishing, and by gathering the edible parts of plants. After the domestication of plants and animals, people were able to remain in one place instead of moving around. **Subsistence agriculture**, in which everything that is produced is consumed, began to replace hunting and gathering. In the 20th century, the dominant form of agriculture in the core countries is **commercial agriculture**, in which crops and animals are produced primarily for sale, and not for consumption by the farmers. See the Agricultural Atlas of the United States website, linked through the textbook's website, for information about the state of agriculture in the United States.

Shifting cultivation, usually found in tropical forests, is a form of agriculture in which farmers aim to maintain soil fertility by rotating the *fields* in which cultivation occurs. Shifting cultivation contrasts with another method of maintaining soil fertility, **crop rotation**, in which the same field is used, but different *crops* are planted in rotation to balance the types of nutrients withdrawn and delivered to the soil. Shifting cultivation is found throughout tropical regions where soil is generally poor, and requires less energy than modern forms of farming even though it can only support small numbers of people. The typical method for preparing a new field is through slash-and-burn, in which the existing plants are hacked close to the ground, left to dry, and then set on fire. This process clears the land and adds new nutrients to the soil. When the field is ready for cultivation, it is known as **swidden**.

Different crops may be grown in different fields or the seeds may be mixed in the same field, which is known as **intertillage**. Intertillage allows for the cultivation of a variety of plants in a small area and it also means that the harvesting of the different plants is staggered throughout the growing season. Another form of subsistence activity is **intensive subsistence agriculture**, a practice that involves the effective and efficient use—through human labor and fertilizer—of a small parcel of land in order to maximize crop yield. This form of agriculture can support much larger populations and is especially characteristic of agriculture in China, India, and Southeast Asia. In some regions, land can be **double cropped**, meaning that more than one crop can be grown in a given year on the same land.

Pastoralism is a third form of agricultural production. **Pastoralism** is a subsistence activity that involves the breeding and herding of animals to satisfy the human needs of food, shelter, and clothing. Pastoralism is often practiced in areas—such as deserts, grasslands, and steppes—where other forms of subsistence agriculture are impracticable. Pastoralism is especially common in parts of North Africa, the Middle East, central Asia, and the Subarctic, and the animals herded are often cattle, sheep, goats, camels, or reindeer, depending on the location. Pastoralists are nomadic, that is, they move from place to place. The movement of herds based on seasonal changes is called **transhumance**.

Agricultural Revolution and Industrialization

Agriculture, like manufacturing, has moved through distinct phases, moving from predominantly subsistence agriculture and towards commercial agriculture. The history of agriculture is characterized by long phases with little change, separated by short phases containing dramatic changes. Three revolutionary periods characterized the development of agriculture: 1) the *First Agricultural Revolution*: beginning before 10,000 BC in Europe and Southeast Asia and characterized by the development of seed agriculture and the use of plow and draft animals, it allowed for the development of settlements. Farming replaced hunting and gathering, and population increased as the land can support more people; 2) the *Second Agricultural Revolution*: beginning around 1650 AD in Western Europe and North America, this revolution is characterized by the production of an agricultural surplus and the development of commercial agriculture, in which the surplus is sold for profit. The second agricultural revolution was closely linked to the Industrial

Revolution taking place at the same time and in the same places; and 3) the *Third Agricultural Revolution*: beginning in 1928 and characterized by the development of agriculture as an industry with industrial methods and policies of production. The emphasis on profit replaces the emphasis on the agrarian way of life, and farms become large commercial enterprises or agribusinesses. This revolution is further characterized by **mechanization**, in which machines replace human labor, by **chemical farming**, in which inorganic fertilizers are applied to the soil to increase yields, and by **food manufacturing**, in which agriculture is linked to the processing and refining of foods. These three revolutions have shaped methods of farming and food production throughout history, leading to the present-day industrialization of agriculture.

Agricultural industrialization is a process in which the role of the farm is moved from being the centerpiece of agricultural production into being only one part of a system of production, storage, processing, distribution, marketing, and retailing of foods. With agricultural industrialization, the farm becomes only one link in a large chain of food production. The process of agricultural industrialization involves three elements: 1) changes in rural labor activities as machines replace and/or improve human labor; 2) the introduction of innovative inputs—fertilizers, hybrid seeds, agrochemicals, and biotechnologies—to supplement, alter, or replace biological outputs; and 3) the development of industrial substitutes for agricultural products (Nutrasweet instead of sugar, and thickeners instead of cornstarch or flour, for example). Agricultural industrialization has not occurred everywhere in the world simultaneously. This process occurred much earlier in the core countries, and was later diffused to the periphery in a process known as the **Green Revolution**, in which technological innovations were exported to the periphery to increase crop yields.

The globalization of agriculture has been accomplished through the same kinds of political and economic restructuring that characterized the globalization of industry. The new ways in which agriculture has been restructured have generated competition, and occasionally conflict, within cultural systems. In many ways, the globalization of agriculture and the role of U.S. farming within it have engendered a farm crisis. The international grain market collapsed in the 1980s; grain prices fell, and mainly American farmers went bankrupt. In response, states such as Nebraska encouraged industries such as meatpacking to establish themselves in the state in the 1990s.

Biological organisms have been used and manipulated in the service of agriculture. An important recent development in this area is **biotechnology**, in which living organisms (or parts of them) are used to modify products, improve plants and animals, or to develop microorganisms for specific uses. Recombinant DNA, tissue culture, cell fusion, enzyme and fermentation technology, embryo transfer, and cloning are some examples of the application of biotechnology. Take a look at the Biotechnology in the Twentieth Century website, linked through the textbook's website, for more information on biotechnology. While biotechnology may lead to many improvements in agricultural efficiency, it can also have negative effects such as the reduced resistance of cloned plants to diseases. Biotechnological developments can also exacerbate core-periphery differences, for example, when plants are developed that can be grown outside their native areas. Private companies normally patent biotechnological innovations, which means that the new technologies are not always widely available.

Global Change in Food Production and Consumption

Globalized agriculture refers to a system of food production increasingly dependent upon the global economy and on international regulatory practices. Today the technologies, political systems, and trade and finance systems that agriculture depends on are global in nature. Agriculture is now part of a larger, more general world economic system (the What Is Sustainable Agriculture? website, linked through the textbook's website, contains a number of further links that explore the implications of agriculture in the world economic system). International and national organizations and agencies play a role in world agriculture. For example, many countries have established agricultural policies that attempt to keep output high, and prices low, for consumers, often through subsidizing production—for example, by establishing artificial prices or by paying farmers to grow or not to grow certain crops. Sometimes these policies result in overproduction and surplus. Governments may attempt to get rid of this surplus by dumping it—selling it below cost—in the international market, or by selling it or donating it to countries that cannot produce enough food.

Redistribution policies resulting from overproduction in core countries may have both positive and negative effects on food recipients. Sometimes, for example, local farmers in the recipient countries are displaced as donated food enters the country, and sometimes food aid reaches only selected segments of the population. Different policies can shape agricultural productivity in a region. In Latin America, **land reform**, in which land was redistributed by the state in the form of **ejidos**, reduced the risk of popular uprisings and helped increase productivity of crops for local consumption. In other parts of Latin America, **aquaculture**, in which fish and shellfish are cultivated, usually in coastal lagoons, has helped increase food production.

Agribusiness is a system that organizes food production from the development of seeds to the retailing and consumption of the agricultural product. Agribusiness is part of the **food chain**, which consists of five central and connected sectors (inputs, production, outputs, distribution, and consumption) and four mediating forces (the state, international trade, the physical environment, and credit and finance). Figure 8.22 in the textbook illustrates connections between the five sectors and the four mediating forces. A **food regime** is a set of links that exist between food production and consumption, and capital investment and accumulation opportunities. The food regime indicates the particular types of food items that are dominant in a given period. A wheat and livestock food regime was in place in the world until the 1960s. Up until this time, wheat and livestock dominated food production, and these products received the bulk of investment. Since the 1960s, vegetables and fruits have become dominant, and the food regime has shifted in this direction. More investment has been directed at developing and marketing vegetables and fruits, and new cultural messages instituted regarding the consumption of these products.

An additional aspect of food production practices that has gained influence in the past 25 years is alternative food movements, including **organic farming**, **local food**, and **slow food** movements. These movements have often emerged in opposition to **fast food**, which has gained dominance in food practices in the United States and elsewhere. In the United States in 2007, 70 billion fast food meals or snacks were served, and the fast food industry employs 13 million people. Every day, one in four Americans eats at a fast food restaurant. The dominance of fast food in the American diet has led to concern about health, nutrition, and obesity.

The Environment and Agricultural Industrialization

Management of the environment is an important part of agriculture. Soil, terrain, water, weather, and insect pests will have an impact on agricultural practices. (The textbook's website will link you to the Agriculture and the Environment website for more information on these issues.) And, conversely, agricultural practices have an impact on the environment. Farming techniques involving the use of pesticides can lead to the presence of toxic chemicals in soil, air, and water, while other farming techniques can lead to a loss of topsoil and to erosion. Especially at risk are lands in arid or semiarid regions, which, through poor management of the environment, can become desert-like, a process known as desertification. Concern for the destruction of some particularly sensitive environments, such as tropical rainforests, has led to economic programs, called debt-for-nature swaps, in which tropical countries of the periphery agree to preserve sections of tropical rainforest in return for a reduction in their foreign debt.

Problems and Prospects in the Global Food System

At present there are two problematic issues in the world food system. The first of these is access to food resources. In some parts of the world, chronic hunger, or **undernutrition**, is a problem. Hunger can also be acute, in the form of **famine**, which is usually short-term but leads to a sharp increase in mortality. A second major problem is the increasing use of genetically modified organisms (GMOs), in which the DNA of an organism is modifed in a laboratory rather than through evolution. Some of the applications of GMOs may be beneficial, but others may be harmful.

Although agriculture is typically thought of as a rural practice, **urban agriculture**, which takes place within or near a city, is also important. An increase in urban agriculture may help improve the food security of people living in cities.

Agriculture has become a highly complex, globally integrated system. Agriculture has become increasingly industrialized through mechanization, through the application of chemical fertilizers, and through the linking of agriculture to the manufacturing, service, and finance sectors of the economy. These changes have affected producers and consumers in both the core and the periphery.

Review Questions

Some of the answers to these questions may be found only in the textbook, and not in the study guide.

1. The cultivation of fish and shellfish under controlled conditions is known as _____.

2. _____ refers to land given to landless peasants in Latin America.

3. Acute starvation associated with a sharp increase in mortality is:
    a) ejidos
    b) famine
    c) undernutrition
    d) transhumance

4. Assured access to enough food at all times to ensure active and healthy lives is called _____.

5. True or False: A genetically modified organism (GMO) is an organism that has had its DNA modified in a laboratory rather than through other forms of evolution.

6. Redistribution of land by the state with a goal of increasing productivity and reducing social unrest is known as _____.

7. The term _____ refers to the culture of agricultural communities and the type of tenure system that determines access to land and the kind of cultivation practices employed there.

8. Three characteristics of the third agricultural revolution are:

_____

_____

_____

9. The set of economic and political relationships that organizes agro-food production from the development of seeds to the retailing and consumption of the agricultural product is called:

    a) intertillage

    b) urban agriculture

    c) food regime

    d) agribusiness

10. The process whereby the farm has moved from being the centerpiece of agricultural production to become one part of an integrated string of vertically organized industrial processes including production, storage, processing, distribution, marketing, and retailing is called _____.

11. _____ is the technique that uses living organisms (or parts of organisms) to make or modify products, to improve plants and animals, or to develop microorganisms for specific uses.

12. True or False: Chemical farming is the application of *organic* fertilizers to the soil and herbicides, fungicides, and pesticides to crops in order to enhance yields.

13. Shifting cultivation is a form of agriculture usually found in _____.

14. True or False: Commercial agriculture is farming primarily for sale, not direct consumption.

15. _____ is the method of maintaining soil fertility where the fields under cultivation remain the same, but the crop being planted is changed.

16. The practice in the milder climates where intensive subsistence fields are planted and harvested more than once a year is called _____.

17. The _____ consists of five central and connected sectors (inputs, production, outputs, distribution, and consumption) with four contextual elements acting as external mediating forces (the state, international trade, the physical environment, and credit and finance).

18. Adding value to agricultural products through a range of treatments—processing, canning, refining, packing, packaging, etc.—that occur off the farm and before they reach the market is known as:
    a) food manufacturing
    b) urban agriculture
    c) agribusiness
    d) food regime

19. Intensive subsistence agriculture is most widespread on the continent of _____.

20. The specific set of links that exists among food production and consumption and capital investment and accumulation opportunities comprise a _____.

21. _____ is a system of food production increasingly dependent upon an economy and set of regulatory practices that are global in scope and organization.

22. The green revolution consisted of an export of fertilizers and high-yielding seeds, from the core to the periphery, in order to:

_____

_____

23. The invention and diffusion of new machines and institutions, from the core to the periphery, to increase global agricultural productivity, is called the _____.

24. Activities whereby people feed themselves through killing wild animals and fish and gathering fruits, roots, nuts, and other edible plants to sustain themselves is called:

    a) intertillage

    b) hunting and gathering

    c) swidden

    d) transhumance

25. _____ is the practice that involves the effective and efficient use—usually through a considerable expenditure of human labor and application of fertilizer—of a small parcel of land in order to maximize crop yield.

26. The farm crisis in the U.S. Midwest in the 1980s was largely caused by a collapse in:

_____

27. _____ is the term applied to the replacement of human farm labor with machines.

28. The subsistence activity that involves the breeding and herding of animals to satisfy the human needs of food, shelter, and clothing is called:

    a) swidden

    b) pastoralism

134 Chapter 8

    c) ejidos

    d) hunting and gathering

29. _____ is a system in which farmers aim to maintain soil fertility by rotating the fields within which cultivation occurs.

30. True or False: Subsistence agriculture refers to farming for direct consumption by the producers, not for sale.

31. The practice of mixing different seeds and seedlings in the same swidden is:

    a) transhumance

    b) aquaculture

    c) intertillage

    d) double cropping

32. Land that is cleared through slash-and-burn and is ready for cultivation is called _____.

33. The movement of herds according to seasonal rhythms—warmer, lowland areas in the winter; cooler, highland areas in the summer—is called:

    a) hunting and gathering

    b) swidden

    c) pastoralism

    d) transhumance

34. Farming for direct consumption by the producers, as opposed to farming for sale, is known as _____ agriculture.

35. Recombinant DNA techniques, tissue culture, cell fusion, and embryo transfer are examples of _____.

© 2010 Pearson Education, Inc.

Conceptual Questions

1. What are the differences between subsistence and commercial agriculture? What regions of the world tend to practice these two basic agricultural modes, and why?

2. What three revolutions took places in agriculture, and what was the impact of each?

3. What is meant by the industrialization of agriculture? What factors account for increasing agricultural industrialization?

136 Chapter 8

_____
_____
_____
_____
_____
_____
_____
_____
_____
_____

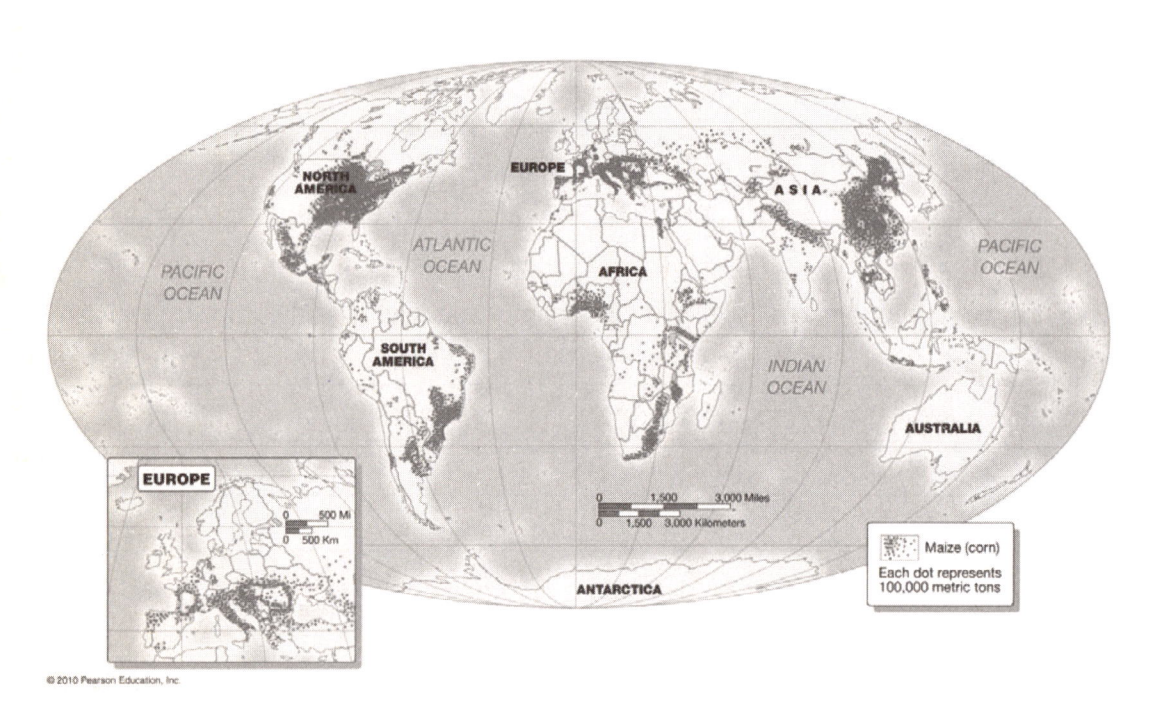

Figure 8.E

4. Using Figure 8.E, which shows the global distribution of maize (corn) production, discuss why maize production is concentrated where it is. Where is the hearth region of maize (it's not indicated on the map)? Why and how did maize diffuse to other regions?

_____
_____

5. What is the Green Revolution? What positive and negative impacts did this process have?

6. What are the five major factors in the food supply system, and how are they connected to each other?

7. What impact has the development of biotechnology had on agriculture? Have the benefits outweighed the costs?

8. What impacts has the industrialization of agriculture had on the environment?

9. How does your own community fit into the food supply system or food web? Is it a producer, distributor, or consumer of agricultural products, or perhaps a combination of these factors? How does this affect the economy of your community?

_____
_____
_____
_____
_____
_____
_____
_____
_____
_____

9

The Politics of Territory and Space

Chapter Objectives

The objectives of this chapter are to:

1. Understand the geopolitical model of the state, and to explore its boundaries and frontiers
2. Examine geopolitics and the world order
3. Prepare a foundation for the understanding of geopolitics

Chapter Notes

The Development of Political Geography

Political geography is a subfield within the discipline of geography. Perhaps the best known area of study within political geography is **geopolitics**, which refers to the study of a state beyond its borders. More specifically, geopolitics refers to the state's power to control space or territory and shape the foreign policy of individual states and international political relations. (Remember that the word states in this case refers to countries, not subdivisions of the United States.) Geopolitics as an area of study developed in 19th-century Germany and was especially influenced by the German geographer Friedrich Ratzel (1844–1904). Ratzel used biological metaphors to describe the state as well as seven laws of state growth. Ratzel's model uses organic metaphors: the state is seen as being like an organism, and, like an organism, it can grow and expand.

The delimited area over which a state exercises control and which is recognized by other states is its **territory**. Territory is enclosed by boundaries. Though boundaries may be fluid and changing, they tend to reinforce spatial differentiation. Different sets of rules may apply in the states on either side of a boundary, and boundaries may limit contact or communication between people on either side of it. Take a look at the International Boundaries Research Unit website, linked through the textbook's website, for information about boundaries and boundary disputes. Frontier regions are places where boundaries are very weakly developed. The United States had a large frontier region throughout much of its history, though today there are few true frontier areas left in the world—Antarctica is one of them. Formal boundaries may be based on natural barriers such as rivers, mountain ranges, lakes, and oceans. Where no natural features occur, or even where they do, boundaries may be drawn by straight lines. Straight lines are very practical: they are easy to survey and easy to delimit on maps of territory that has not been fully explored or settled. Internal boundaries—those within countries—are also drawn on the same basis. Territories delimited by formal boundaries are known as *de jure* (meaning legally recognized) spaces or regions. These spaces, such as countries, states of the United States, counties, municipalities, special districts, and so forth, are the basic units of analysis in human geography and other social sciences, as they are normally the only units for which data can be obtained.

Geopolitics and the World Order

A state is an independent political unit with recognized boundaries. (You can get detailed information about the world's states through the textbook's website link to the World Governments site.) In contrast to a state, a **nation** is a group of people sharing common elements of culture such as religion, language, history, or political identity. A **nation-state** is an ideal form consisting of a homogeneous group governed by their own state. Very few states are true nation-states, because most states contain minority peoples who belong to another nation. Denmark, however, is a close approximation of a nation-state: the territory of Denmark corresponds almost exactly with the area inhabited by ethnic Danes. **Sovereignty** is the exercise of state power over people and territory recognized by other states and by its own people. Most states today are multinational; that is, they contain people from more than one nation or cultural group. Sometimes the nations composing a multinational state live peacefully together, but often there are tensions and occasionally major conflicts. **Nationalism** is the feeling of belonging to a nation as well as the belief that a nation has a natural right to determine its own affairs. The province of Québec in Canada is a good example of tensions in a multinational state and also of nationalism: many of the French-speaking residents of Québec consider themselves as part of a distinctive nation and have at times agitated for separation from Canada. Also see the Tibet Independence Movement and Scottish Independence Movement websites, linked through the textbook's website, for examples of other nationalist movements.

Both centripetal and centrifugal forces beset multinational states. **Centripetal forces** are those that strengthen and unify the state. For example, during the era of the Soviet Union, national differences, such as Russian, Ukrainian, Estonian, and Kazakh, were downplayed in favor of an umbrella identity known as Soviet Man. The creation of a Soviet identity was a centripetal force that attempted to hold the multinational Soviet Union together. **Centrifugal forces** are those that divide or tend to pull the state apart. Nationalism among the various nations comprising the Soviet Union was a centrifugal force, and one that ultimately pulled the country apart. One way in which multinational states have attempted to defuse potential tensions is by establishing a federal state. A **federal state** is a form of government in which power is allocated to units of local government within the country. The United States is a federal state, with power shared between the national, or federal, government and the state governments. Other examples of federal states include Canada, Australia, Germany, and Switzerland; and, to a more limited extent, India and the former Soviet Union. Federal states can be contrasted with **unitary states**, which are a form of government in which power is concentrated in the central government. Examples of unitary states include Sweden and France.

The breakup of the former Union of Soviet Socialist Republics, or Soviet Union, provides a good example of the problems of multinational states. The Soviet Union, created after the Russian Revolution of 1917, inherited a territory from the former Russian Empire that was ethnically diverse, with over 200 different national groups. The larger nations within the Soviet Union were granted limited autonomy under a federal system, and were referred to as Soviet Socialist Republics. When Mikhail Gorbachev came to power in the Soviet Union in 1985, he attempted to reform the existing Soviet system through policies of *perestroiyka* (restructuring) and *glasnost* (openness). Gorbachev found allies for his reform efforts in politicians from the national republics, but, by lifting restrictions on national organization, he encouraged nationalism that eventually led to the breakup of the country. Most of the former Soviet republics are still loosely organized as a confederation, called the Commonwealth of Independent States (CIS). A **confederation** is a group of states united for a common purpose. Another example of a confederation was the Confederate States of America, comprised of the eleven southern states that seceded from the United States between 1860 and 1861. This secession led ultimately to the Civil War. While the breakup of the Soviet Union was relatively peaceful, the shifting of state boundaries in the Balkans resulted in bitter conflicts. Yugoslavia was a state composed of many nations; its dissolution, in the 1990s, led to war between several of these constituent nations, and conflict still simmers in the Kosovo region.

Political geographers have constructed theories explaining state behavior. A state is not only a geographical place; it is also a *set of institutions* for the protection and maintenance of society. The geopolitical theories of Friedrich Ratzel, discussed above, constitute a theory of the state, using the metaphor of an organism. Geopolitics was later adapted and corrupted by German political geographers during the Nazi regime, who used it as a justification for exterminating non-Germans and for invading neighboring states. Imperialism and colonialism are two ways in which states can exercise power and control over external territories. Colonialism involves formal establishment of control over a foreign population through settlement, while imperialism does not necessarily imply formal control but rather pressures such as military threat, economic sanctions, or cultural domination. The process of imperialism begins with exploration, such as that which characterized European expansion beginning in the 15th century.

Colonialism is responsible for many of the problems of contemporary multinational states. Often colonizing powers—notably European powers in Africa—divided up colonial territories with little thought to the boundaries between existing nations. Thus European-drawn boundaries in colonial Africa were often straight lines, sometimes dividing one African nation between two colonies, and sometimes placing two hostile African nations within the same colony. When these colonies became independent, the colonial boundaries were maintained, leading to some of the conflicts seen within and between African countries today. You can take a look at some of these conflicts by going to the Canadian Department of National Defence site on current wars and conflicts, linked through the textbook's website. Another legacy of colonialism is the **North/South divide**, or differentiation made between the colonizing states of the Northern Hemisphere and the formerly colonized states of the Southern Hemisphere. This divide is characterized by a relation of dependence, in which the countries of the South are economically dependent on countries of the North.

Decolonization is the acquisition, by colonized peoples, of control over their own territory. The United States achieved its independence relatively early, in the 18th century. Most Latin American colonies received their independence in the 19th century, while colonies in Africa and Asia largely received their independence in the 20th century. The massive decolonization that took place in the 20th century was partly set in motion by the colonial powers themselves. Decolonization does not necessarily mean an end to domination within the world system. Many independent former colonies are still politically, economically, and culturally linked to the former colonizing power. The core countries provide development aid and assistance, and also cultural styles and fashions.

Imperialism and colonialism were both supported by geopolitical theories developed in the colonizing countries, and these theories helped Europeans justify their colonial activities. The English geographer Halford Mackinder (1861–1947) developed one such theory—*heartland theory*. His theory argued that Eurasia was the "geographical pivot" or "heartland," a location central to establishing global control, and he divided the world into three regions: heartland, inner crescent, and outer crescent. He also argued that land power was more important than sea power in achieving global control. As Russia controlled the heartland, Mackinder suggested that Russia be prevented from expanding. Many British colonies in Asia and Africa were reinforced as outposts preventing the expansion of Russia, and the theory was used to justify the British presence in India.

In addition to the North/South divide based on imperialism and colonialism, the world order of states can also be seen to cluster along an East/West split. The **East/West**

divide refers to communist and noncommunist countries, respectively. The East/West divide was a significant aspect of the world system between 1945 and 1990. One aspect of the East/West divide reflected in American foreign policy was the **domino theory**. The domino theory argued that if a region chose or was forced to accept a communist political and economic system, then neighboring countries would be irresistibly susceptible to falling to communism. This theory was used to justify U.S. intervention in various places around the world, most notably in Vietnam in the 1960s and 1970s, and is still used to justify an embargo on trade with Cuba. The concept of the **new world order** refers to a belief that, with the triumph of capitalism over communism, the United States becomes the world's only superpower and therefore its policing force. This process, however, has created instability in some parts of the world, and has also meant that traditional forms of warfare and political practices have been replaced by more radical ones. One of these is **terrorism**, in which individuals or groups have taken violent action against civilian populations in order to undermine state practices or institutional organizations.

International and Supranational Organizations and New Regimes of Global Governance

International and supranational organizations have become important participants in the world system in the 20th century. An **international organization** is one that includes two or more states seeking political and/or economic cooperation with each other. Examples include the United Nations (UN) and the Organization of Petroleum Exporting Countries (OPEC). International organizations serve various purposes while maintaining full sovereignty for their individual member states. **Supranational organizations**, such as the European Union (EU), reduce the centrality of individual states, and each member state may have to give up a small portion of its sovereignty as a condition of membership. The International Integration Theory website, linked through the textbook's website, provides more information on supranational organizations. Global politics has become increasingly institutionalized, and economic, ecological, and social issues have assumed greater prominence in international affairs. An **international regime** refers to the international networks generated by this process and to the increasing internationalization of politics.

The Two-Way Street of Politics and Geography

Political geography is about both the *politics of geography* and the *geography of politics*. The politics of geography emphasizes the impact that geography has on politics. The geography of politics emphasizes the impact that politics has on geography. Regionalism and sectionalism illustrate the politics of geography. Both are connected to the concept of **self-determination**, which argues for the right of a group with a distinct political and territorial identity to determine its own destiny, at least in part, through the control of its own territory. **Regionalism** is the feeling of collective identity based on a group's political and territorial history. Regionalism often involves groups seeking autonomy or independence from what they perceive as a foreign and illegitimate state. An example of regionalism is the Basque separatist movement in northern Spain and southern France, which seeks a separate and autonomous Basque homeland. Another example is the current conflict in Northern Ireland, where Irish Catholics seek union with the Republic of Ireland and Irish Protestants prefer to remain within the United Kingdom. Regionalism may also be based on factors other than ethnicity, as in the occasional separatist movements that pop up in western Canada, arguing that the country is dominated by eastern interests that neglect the needs of the Canadian West.

Sectionalism is a different concept, and refers to the extreme devotion to local interests and customs. Sectionalism has been identified as an overarching explanation for the United States Civil War. Attachment to slavery and the way of life that slavery enabled led to the secession of eleven southern states. Sectionalism in the United States persists today, but largely in the form of conflicts between American cities and their suburbs, and the ways of life that these two municipal units represent.

Systems of political representation are geographically anchored, and reflect the geography of politics. **Territorial organization** is a system of government structured by representation by area, not by social groups. Thus representation in the United States Congress and state legislatures consists of persons representing territories, such as states or congressional districts, and not social groups such as labor, industry, agriculture, and so forth. Within the United States there are a variety of different electoral divisions, each based on territory. Thus, in order for individuals and their interests to gain representation, they must be able to capture control of *geographically based* political units. In other words, if farmers want representation in Congress, they must gain control of geographically based political units such as states or congressional districts. The United States presidential electoral system is likewise geographically based. Presidential candidates must gain a majority of votes in the electoral college, based largely on state boundaries. In order to win the electoral college votes for a given state, the candidate must win a plurality of votes in that state. Thus presidential candidates usually concentrate their campaigning in the most populous states where, by winning a plurality of votes, they can gain large numbers of electoral votes.

The number of congressional districts—and thus the number of congressional representatives—from each state is adjusted after each census. States with increased population may gain congressional seats, while states with declining, stable, or slowly growing populations may lose representation. The process of allocating electoral seats to geographical areas is called **reapportionment**. **Redistricting** is the process of defining and redefining territorial district boundaries, as must happen, for example, if a state increases its number of congressional seats. The intent of redistricting is to insure the equal probability of representation among all groups. An abuse of this practice, in which districts are drawn to serve the ends of political parties, is known as **gerrymandering**. In this practice, districts may be redrawn in unusual ways resulting in unusual shapes, in order to insure that a political party will control the district. Figure 9.36 in the textbook illustrates one such district.

States are the building blocks of the global political order, and the globalization of the economy has been facilitated by states acting beyond their boundaries. The processes of colonization and decolonization have resulted in tensions and conflicts throughout the world and have helped to solidify the North/South divide. As this chapter has shown, politics is geographical, and geography can be political.

Review Questions

Some of the answers to these questions may be found only in the textbook, and not in the study guide.

1. The rebellion, or uprising, among the Palestinians in the 1980s and since then is known as _____.

2. The notion of the new world order assumes that _____ becomes the world's only superpower.

3. Give three examples of terrorism:
_____
_____
_____

4. A feeling of collective identity based on a population's politico-territorial identification with a state or across state boundaries is known as _____.

5. _____ was a 19th-century movement that eventually led to the formation of the Jewish state of Israel.

6. _____ is the defining and redefining of territorial district boundaries.

7. Today, domino theory and heartland theory have largely been replaced by theories of:
_____

8. True or False: Centrifugal forces: forces that strengthen and unify the state.

9. A group of states united for a common purpose is a:

    a) confederation

    b) federal state

    c) unitary state

    d) international organization

10. _____ is the acquisition, by colonized peoples, of control over their own territory.

11. A _____ region occurs where boundaries are very weakly developed.

12. The notion that if one country in a region chose or was forced to accept a communist political and economic system, then neighboring countries would be irresistibly susceptible to falling to communism, was known as:

    a) North/South divide

    b) geopolitics

    c) Zionism

    d) domino theory

13. _____ and _____ are two Southern Hemispheric states that are part of the North in an economic sense.

14. The _____ refers to communist and non-communist countries respectively.

15. Two ways in which the geopolitical extension of power by one group over another can occur are:

_____

_____

16. The Basque separatist movement in Europe is an example of _____.

17. A form of government in which power is allocated to units of local government within the country is called a:

    a) confederation

    b) international regime

    c) federal state

    d) unitary state

18. The practice of redistricting for partisan purposes is called:

    a) reapportionment

    b) redistricting

    c) gerrymandering

    d) geopolitics

19. A group that includes two or more states seeking political and/or economic cooperation with each other is called an _____.

20. True or False: A state is a group of people often sharing common elements of culture such as religion or language, or a history or political identity.

21. _____ is an ideal form consisting of a homogeneous group of people governed by their own state.

22. True or False: Centripetal forces are forces that divide or tend to pull the state apart.

23. The state's power to control space or territory and shape the foreign policy of individual states and international political relations is called _____.

24. The feeling of belonging to a nation as well as the belief that a nation has a natural right to determine its own affairs is called _____.

25. True or False: The North/South divide refers to the differentiation made between the colonizing states of the northern hemisphere and the formerly colonized states of the southern hemisphere.

26. _____ is the process of allocating electoral seats to geographical areas.

27. The right of a group with a distinctive politico-territorial identity to determine its own destiny, at least in part, through the control of its own territory is called:
   a) self-determination
   b) international regime
   c) sectionalism
   d) regionalism

28. The exercise of state power over people and territory, recognized by other states and codified by international law, is called _____.

29. _____ is a system of government formally structured by area, not by social groups.

30. True or False: A nation is the delimited area over which a state, an individual, or a group exercises control and which is recognized by other states, individuals, or groups.

31. A form of government in which power is concentrated in the central government is called a:
   a) confederation
   b) federal state
   c) unitary state

d) colony

32. Forces that divide or tend to pull the state apart are _____ forces.

33. A _____ state is one in which power is allocated to units of local government within the country.

34. True or False: Sectionalism refers to extreme devotion to local interests and customs.

35. _____ is a system in which public policies and officials are directly chosen by popular vote.

36. The 19th-century German geographer Friedrich Ratzel suggested that states, like living organisms, progress through the three stages of:
_____

37. A _____ is a group of people sharing common elements of culture such as religion or language, or a history or political identity.

Conceptual Questions

1. What is the difference between a nation and a state? Can you give some examples of nation-states as well as multinational states?
_____
_____
_____
_____
_____
_____

2. Are there any frontier regions left in the world today? What are some of the problems that these regions face?

3. What kinds of boundaries does your state of residence (national and sub-national state) have? Why were boundaries drawn in these particular ways?

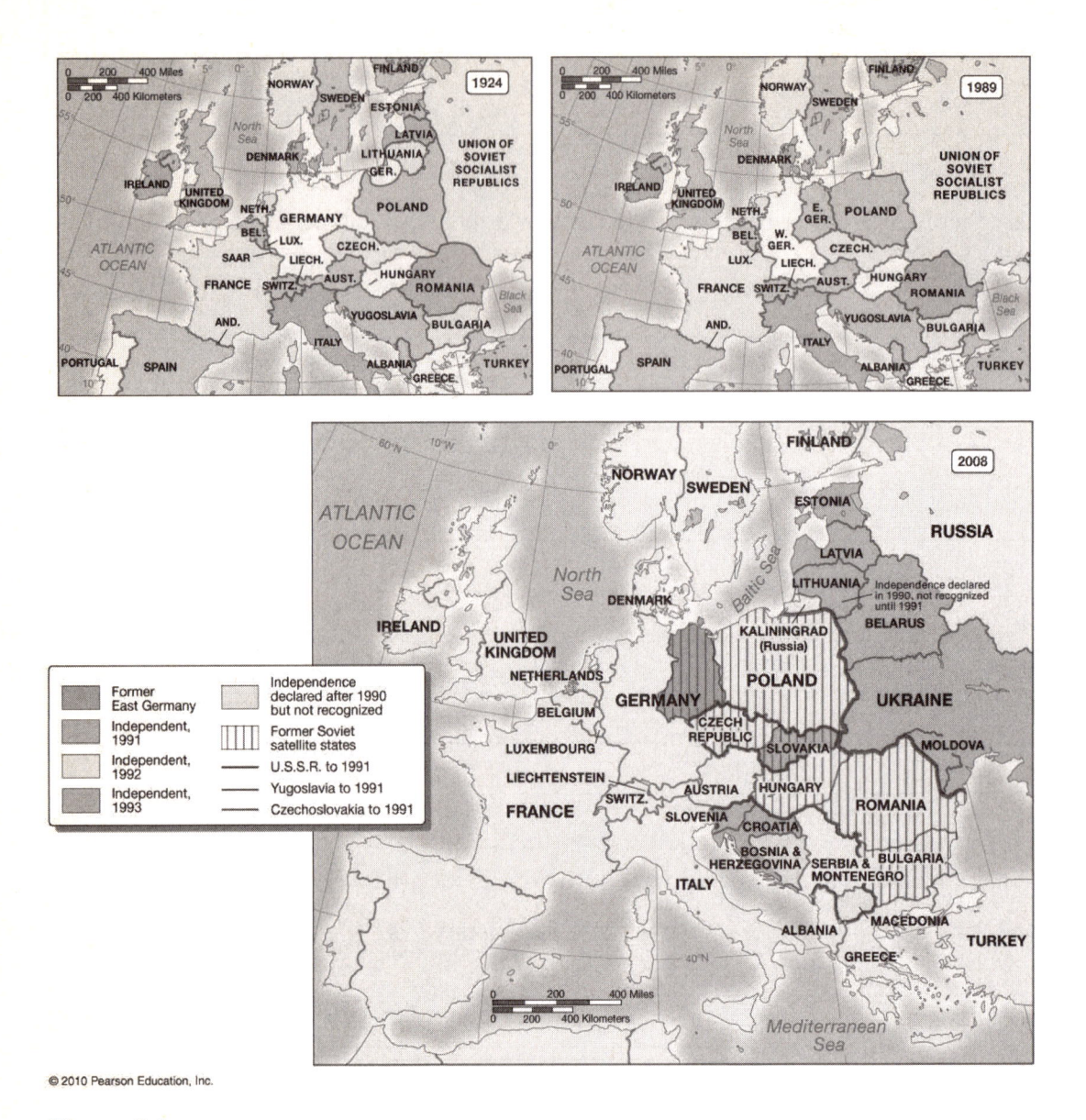

Figure 9.1

4. We tend to think of Europe as a place with stable political boundaries. Yet Figure 9.1 shows how dramatically European boundaries have changed during the past hundred years. What has accounted for these changes?

_____
_____
_____
_____

5. Give some examples of nationalist movements in the world today. What are these movements trying to achieve? How are they going about achieving it?

6. What is meant by the North/South divide? How did this divide come about and what implications does it have for the world of today?

156   Chapter 9

7. What is domino theory, and how did it feature in U.S. foreign policy? Is it still a feature of U.S. foreign policy today?

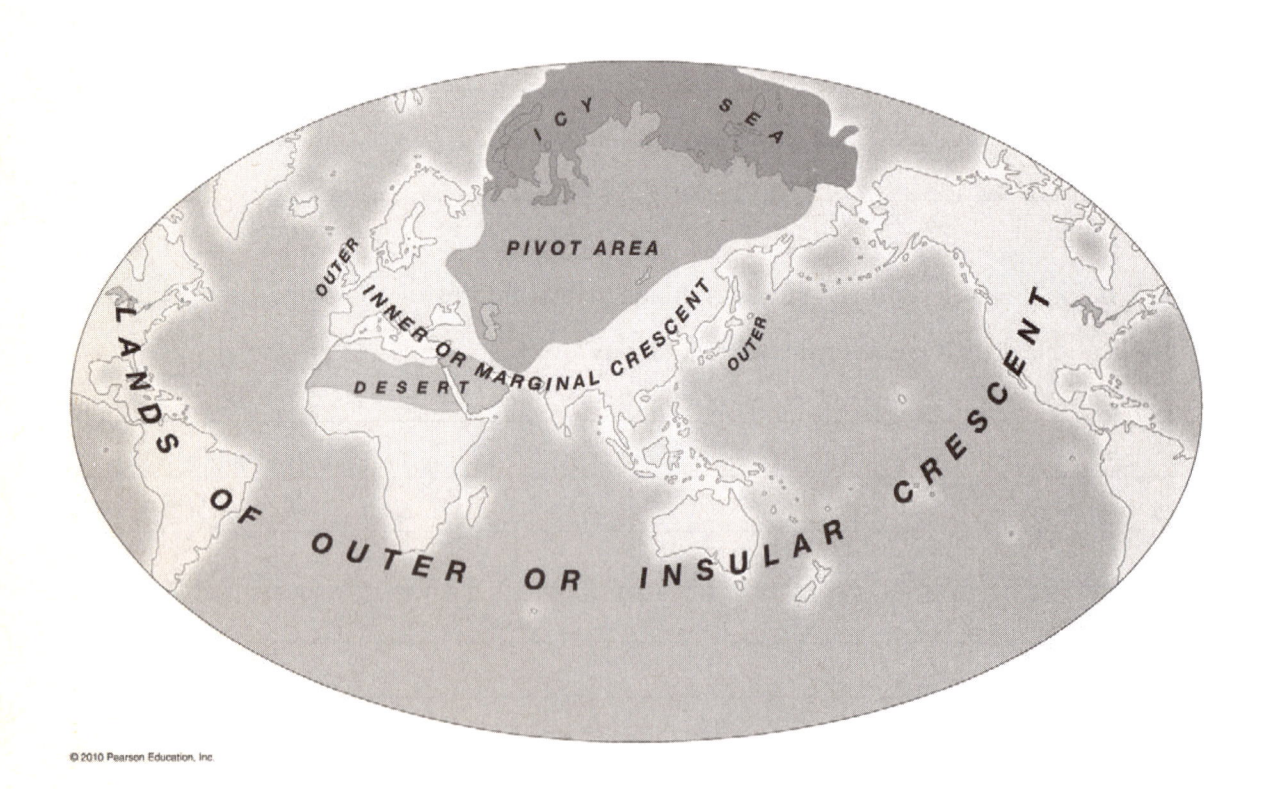

Figure 9.24

8. Figure 9.24 illustrates Halford Mackinder's theory of the heartland. Explain the basics of the heartland theory. Do you think this theory is correct? How were maps like these used for political purposes?

9. What are the boundaries of your federal congressional district? How was this boundary drawn? Is it a gerrymandered boundary? Why or why not?

10

Urbanization

Chapter Objectives

The objectives of this chapter are to:

1. Examine the roots of European urban expansion
2. Explore today's urbanization, looking at regional trends and projections, and their urban systems
3. Investigate urban growth processes

Chapter Notes

Urban Geography and Urbanization

Urban communities have always been an important element in spatial organization, but they are now more important than ever before. Cities now account for almost half of the world's population, and much of the developed world has become almost completely urbanized. Towns and cities are engines of economic development and centers of cultural innovation, social transformation, and political change. Towns and cities have four key roles: 1) the *mobilizing function* of urban settlement. Cities provide efficient and effective environments for organizing labor, capital, and raw materials, and for distributing finished products; 2) the *decision-making capacity* of urban settlements. Cities bring together decision-makers in public and private institutions, and thus become concentrations of political and economic power; 3) the *generative functions* of urban settlement. The concentration of people in urban areas brings about greater interaction and competition, leading to innovations, knowledge, and information; and 4) the *transformative capacity* of urban settlement. The size and variety of cities allows them to have a liberating effect on people, freeing them from the rigidities of traditional, rural society. City environments allow people to engage in a variety of lifestyles and behaviors.

Urban geographers study these functions of urban settlements, and examine the similarities and differences both among and within urban places around the world. An **urban system**, or city system, is an interdependent set of urban settlements within a given region. We can thus speak of the French urban system, the African urban system, or the global urban system, for example. **Urban form** is the physical structure and organization of cities in terms of their land use, layout, and built environment. Cities not only change in size, but also get reorganized, redeveloped, and redesigned. **Urban ecology** is the social and demographic composition of city districts and neighborhoods. The numbers and varieties of people within cities changes, and different city neighborhoods and districts will reflect these changes. (You can take a look at comprehensive data about your community from the U.S. census by going to the U.S. Gazetteer website, linked through the textbook's website.) The way of life, attitudes, and patterns of behavior fostered by urban environments is known as **urbanism**. Urbanism refers to what makes life in cities different from life in small towns or in rural areas.

Urban Origins

Many cities are the results of long periods of development. In order to understand cities, we need to examine the reasons for their growth, their rates of growth, and the factors that have contributed to that growth. The earliest urban settlements emerged in the hearth areas of the first agricultural revolution, especially in Mesopotamia and the Nile Valley. Urbanization in these areas may have resulted from the presence of an agricultural surplus, which was large enough to allow for the emergence of non-agricultural specialists, or it may have resulted from population pressures, as people had to develop new forms of economy. Urbanization requires an elite group who can impose taxes and control labor for the benefit of the city as a whole. This elite financed and constructed large public buildings and monuments. The emergence of cities also created specialized occupations such as construction, crafts, administration, the military, and so forth.

In Europe, the Greeks and Romans had established organized urban systems. Though the urban system of Europe declined in the early Middle Ages (5th to 10th centuries), it re-emerged around the 11th century. The early Middle Ages were characterized by a system of *feudalism*, based upon self-sufficient rural estates. As population grew, the feudal system began to collapse, replaced by a system emphasizing a money economy and trade. This was the basis for a new phase of urbanization in Europe beginning in the 11th century. Take a look at some of the maps of ancient and medieval cities at the Interactive Historical Urbanization Maps website, linked through the textbook's website.

The emerging urbanism of the 11th century was based on a new form of economy known as merchant capitalism. Merchant capitalism involved both trade and the accumulation of capital for further investment. Cities such as Venice, Florence, Hamburg, and Amsterdam grew as large trading centers during this period. A complex network of trade soon covered most of Europe. A series of transformational changes took place in Europe in the 15th and 16th centuries. Merchant capitalism became more sophisticated, and the Protestant Reformation and the Scientific Revolution stimulated economic and social reorganization. Spanish and Portuguese colonialism followed Spanish and Portuguese exploration in Africa, Asia, and the Americas. The European urban system was gradually extended to these colonized areas. For example, the Spanish established colonial cities on the sites of older indigenous cities or where valuable minerals were found.

In Europe, Renaissance organization saw the centralization of political power and the formation of national states; the beginnings of industrialization; and the funneling of plunder and produce from distant colonies. In this context, the port cities near the Atlantic had an advantage, as they were closer to the maritime trade routes with Africa, Asia, and the Americas. **Gateway cities**, those that serve as a link between one country or region and others because of their physical location, became especially important during this period (see Figure 10.9 in the textbook). Gateway cities of the colonial era included Rio de Janeiro, Boston, Kolkata (Calcutta), and Accra. Each gateway city funneled products from the interior of its region to colonial ships, where the products could be transported to the mother country. The Industrial Revolution and European imperialism were closely linked to urbanization. Industrialization required urbanization, because the demands for labor, capital, transportation, and markets could only be met in cities. Higher wages in industrial and trade cities attracted migrants, increasing the size and influence of these cities.

Cities have a strong influence on the economy and culture of a region. A city that is seen as the embodiment of surprising and disturbing changes in economic, social, and cultural life is known as a **shock city**. **Colonial cities** are those that were deliberately established or developed as administrative or commercial centers by colonial or imperial powers. In some cases colonial cities were built where no settlement previously existed, while in other cases the colonial administrative and commercial functions were added to existing urban settlements. Colonial cities are often based on urban models from the colonizing country. For example, cities in Spanish colonies in the Americas, whether they were built in deserts or tropical regions, followed the same plan, that of a central plaza and a grid-like system of streets. Santa Fe in the United States, Havana in Cuba, and Potosí in Bolivia were all built in this way.

Urban Systems

Urban systems link settlements together, and are organized hierarchically. A **central place** is a settlement in which certain products and services are available to consumers, and **central place theory** attempts to explain the relative size and spacing of towns and cities based on people's shopping behavior. In other words, why are cities located where they are, and why are they the size that they are? There is often a relationship between the size of cities and their rank in the overall hierarchy. This is known as the **rank-size rule**. The relationship is such that the nth largest city in a country or region is $1/n$ the size of the largest city in that country or region. Thus, the fifth largest city would be about one-fifth the size of the largest city. Thus, in the United States, the second largest city, Los Angeles, is about half the size of the largest, New York. In some urban systems, the largest city is disproportionately larger than the rank-size rule would suggest. Buenos Aires, for example, is ten times (not twice) the size of the second largest city in Argentina, Rosario. The same disproportionate pattern is found in such places as the United Kingdom, France, and Brazil. These disproportionately large cities are known as *primate cities*, reflecting the condition of **primacy**. Primacy is often a result of a city's early role as a gateway city.

When a city's economic, political, and cultural functions are disproportionate to its population, the condition is known as **centrality**. Primate cities often exhibit centrality, but a city—such as New York—does not have to be a primate city in order to exhibit centrality. *World cities* are those that play key roles in the global urban system. They possess several characteristics: they are the sites of most of the leading global markets for commodities, commodity futures, investment capital, foreign exchange, equities, and bonds; they are the sites of clusters of specialized, high-order business services, especially those that are international and connected to finance, accounting, advertising, property development, and law; they are the sites of concentrations of corporate headquarters; they are the sites of most of the leading non-governmental organizations and intergovernmental organizations; and they are the sites of the most powerful and internationally influential media organizations, news and information services, and culture industries. Today the world system is dominated by two cities with global influence: London and New York.

Urbanization is a relative phenomenon. What is considered urban in one country may not be considered urban in another. In Canada a settlement with 1500 people is officially classified as urban, while in Japan a settlement with 45,000 is *not* officially classified as urban. These classifications indicate something about the demographics of the country and its perceptions of urban and rural. Taking the definitions used in individual countries, then worldwide statistics indicate that almost one-half of the world's population is urbanized. Using this criterion, North America emerges as the most highly urbanized continent, with 81.3 % of its population in urban areas. Asia and Africa are the least urbanized. Data for all continents is presented in Table 10.1 in the textbook.

Today the countries of the periphery are urbanizing much faster than the countries of the core. In 1950, 21 of the world's 30 largest metropolitan areas were located in core countries, while in 1980 only 11 of the world's 30 largest cities were located in core countries. The cities of the periphery are growing much faster than the cities of the core. The process of urbanization in the world's peripheral regions has been different from that of the core regions. Urbanization in the periphery largely results from population increases and not necessarily from increased agricultural efficiency. As rural residents become increasingly impoverished, many move to cities where they hope to find better jobs. The consequence of this rapid growth is **overurbanization**, in which cities grow more rapidly than the jobs and housing they can sustain. New slums appear, often as **squatter settlements**, in which houses are built on land that is neither owned nor rented by the occupants.

Rapid urban growth has also taken place in frontier regions such as the Brazilian Amazon, where the government has encouraged migration, partly by building roads into the jungle. The Brazilian government encourages this frontier urbanization because it reduces stress in the already overurbanized cities of Rio de Janeiro and São Paulo. In addition to industrialization and growth, cities may also suffer deindustrialization and loss of population. Deindustrialization involves a decline in industrial employment and loss of industry. **Counterurbanization** occurs when cities experience a net loss of population to small towns and rural areas. This happened in the United States, for example, in the 1970s and 1980s, when people left older industrial towns such as Cleveland for newer, smaller communities in the Sunbelt.

Megacities are not necessarily world cities, though some of them are. Megacities are characterized by primacy, a high degree of centrality, and by their very large size. Examples include Bangkok, Beijing, Cairo, and Mexico City. These cities do not only link local and national economies with the global economy; they also link the formal and informal economic sectors. Activities in the **informal sector** take place beyond official record and are not subject to formal systems of regulation or remuneration.

Globalization and Splintering Urbanism

Urbanization in the world's core regions was stimulated by advances in agricultural productivity, which increased the food supply while at the same time reducing the numbers of farmers (as discussed in Chapter 8). Urbanization also led to increased industrial production of farm machinery that in turn further increased agricultural productivity. This cycle allowed industrial cities to grow at a rapid rate. Today many cities show characteristics of **splintering urbanism**, which refers to intense geographical differentiation, with cities and parts of cities engaged in complex and changing economic and technical exchange. World cities and major regional centers have been the primary beneficiaries of splintering urbanism. Cities and urban areas organize not only their own spaces but also those around them: the influence of cities is disproportionately greater than their geographical size. The experiences of urbanization have differed in the core and periphery, such that today the largest cities in the core countries are world cities with global influence, while many in the periphery are overurbanized, struggling to deal with the problems of rapid urbanization.

Review Questions

Some of the answers to these questions may be found only in the textbook, and not in the study guide.

1. In broad terms, the earliest urbanization developed in this region of the world:
_____

2. According to the rank-size rule, the nth-largest city in a country or region is ____/____ the size of the largest city in that country or region.

3. The growth in population in metropolitan central cores, following a period of absolute or relative decline in population, is known as _____.

4. A settlement in which certain products and services are available to customers is called a:

    a) colonial city

b) central place

c) world city

d) gateway city

5. _____ is a statistical regularity in city-size distributions of cities and regions.

6. The physical structure and organization of cities is called _____.

7. _____ is a theory that seeks to explain the relative size and spacing of towns and cities as a function of people's shopping behavior.

8. The world's largest city, by population, in both 1980 and 2010 (projected) is _____.

9. Most of the "world cities" are located in the _____ regions of the world.

10. True or False: Centrality refers to the functional dominance of cities within an urban system.

11. Cities that were deliberately established or developed as administrative or commercial centers by colonial or imperial powers are called _____.

12. _____ was the key factor in fueling the rapid urbanization of Europe in the first half of the 19th century.

13. True or False: Overurbanization is the net loss of population from cities to smaller towns and rural areas.

14. A city that serves as a link between one country or region and others because of its physical situation is called a:

    a) colonial city

    b) gateway city

    c) world city

    d) shock city

15. True or False: The informal sector refers to economic activities that take place beyond official record, not subject to formalized systems of regulation or remuneration.

16. A _____ is a very large city characterized by both primacy and has high centrality within its national economy.

17. Karachi, New York, and Mumbai (Bombay) are all very large cities, but they are not _____ cities.

18. _____ is a condition in which cities grow more rapidly than the jobs and housing they can sustain.

19. The continent of _____ has the greatest percentage of its population in urban areas.

20. The condition in which the population of the largest city in an urban system is disproportionately large in relation to the second- and third-largest cities in that system is called:

    a) centrality

    b) primacy

    c) counterurbanization

    d) overurbanization

21. Give two examples of shock cities:
_____

22. Rio de Janeiro in Brazil, Accra in Ghana, and Buenos Aires in Argentina are all examples of _____ cities.

23. Residential developments that take place on land that is neither owned nor rented by its occupants are called _____.

24. An interdependent set of urban settlements within a specified region is called an:
   a) shock city
   b) urban ecology
   c) urban system
   d) urban form

25. True or False: Urbanism: the way of life, attitudes, values, and patterns of behavior fostered by urban settings.

26. A city in which a disproportionate part of the world's most important business is conducted is called a:
   a) shock city
   b) gateway city
   c) colonial city
   d) world city

27. _____ theory seeks to explain the relative size and spacing of towns and cities as a function of people's shopping behavior.

Conceptual Questions

1. Would you rather live in an urban or rural setting? Explain the reasons for your answer.

2. What four functions do cities and towns serve? How are these reflected in your own community, if it is an urban one?

3. What factors contributed to the growth of European cities in the late Medieval and Renaissance periods?

4. What is meant by a shock city? Can you think of any examples of cities that are considered shock cities today?

5. What were some of the distinctive features of colonial cities? If you live in a colonial city, are any of these features still visible in your community?

6. How has the growth of cities in the periphery been different from the experience of core cities? What distinctive problems do peripheral cities face?

7. What is the nearest world city to where you live? What effects does this city have on your life?

TABLE 10.2 The World's 30 Largest Metropolitan Areas, Ranked by Population Size, 1950, 1980, and 2010 (in millions)

1950	Population	1980	Population	2010	Population
New York	12.3	Tokyo	28.5	Tokyo	36.0
London	8.7	New York	15.6	Mumbai	24.0
Tokyo	6.9	Mexico City	13.0	São Paulo	21.1
Paris	5.4	São Paulo	12.1	Mexico City	20.6
Moscow	5.4	Shanghai	11.7	New York-Newark	19.4
Shanghai	5.3	Osaka	10.0	Delhi	17.0
Essen	5.3	Buenos Aires	9.9	Shanghai	15.7
Buenos Aires	5.0	Los Angeles	9.5	Kolkata	15.6
Chicago	4.9	Kolkata	9.0	Dhaka	14.7
Kolkata	4.4	Beijing	9.0	Buenos Aires	13.0
Osaka	4.1	Paris	8.8	Karachi	13.0
Los Angeles	4.0	Mumbai	8.7	Los Angeles	12.7
Beijing	3.9	Rio de Janeiro	8.6	Cairo	12.5
Milan	3.6	Seoul	8.3	Rio de Janeiro	12.1
Berlin	3.3	Moscow	8.1	Beijing	11.7
Mexico City	3.1	London	7.7	Manila	11.6
Philadelphia	2.9	Cairo	7.3	Osaka-Kobe	11.3
St. Petersburg	2.9	Tianjin	7.3	Lagos	10.6
Mumbai	2.9	Chicago	7.2	Istanbul	10.5
Rio de Janeiro	2.9	Rhein-Ruhr	6.3	Moscow	10.5
Detroit	2.8	Jakarta	6.0	Paris	10.0
Naples	2.8	Metro Manila	6.0	Seoul	9.8
Manchester	2.5	Delhi	5.6	Jakarta	9.7
São Paulo	2.4	Milan	5.3	Guangzhou	9.4
Cairo	2.4	Tehran	5.0	Chicago	9.2
Tianjin	2.4	Karachi	5.0	Kinshasa	9.0
Birmingham	2.3	Bangkok	4.7	London	8.6
Frankfurt	2.3	St. Petersburg	4.6	Lima	8.3
Boston	2.2	Hong Kong	4.6	Bogotá	8.3
Hamburg	2.2	Philadelphia	4.5	Tehran	8.2

Source: Data from United Nations Department of Economic and Social Affairs/Population Division, *World Urbanization Prospects: The 2007 Revision.* New York: UN Department of Economic and Social Affairs, 2008, p. 167.

© 2010 Pearson Education, Inc.

Table 10.2

8. Look at Table 10.2, which shows the world's largest metropolitan regions in 1950 and 1980, and projections for 2010. What trends do you notice? What has accounted for these changes and trends?

11

City Spaces: Urban Structure

Chapter Objectives

The objectives of this chapter are to:

1. Examine urban structure and land use
2. Investigate urban form and design
3. Survey urban trends and problems

Chapter Notes

Urban Land Use and Spatial Organization

Different groups of people may cluster in different parts of the city, a process known as congregation. **Congregation** is especially characteristic of **minority groups**, population subgroups that are seen—and see themselves—as different from the general population. Minority groups may be defined on the basis of ethnicity, language, religion, nationality, caste, sexual orientation, or lifestyle. Congregation helps to preserve and strengthen the group's culture; it helps minimize conflicts with non-group members; it provides a place for mutual support through minority institutions, businesses, social networks, and welfare organizations; and it helps establish a power base for the group. Congregation may be voluntary, or it may be forced as a result of discrimination. Congregation and discrimination together may result in **segregation**, which is the spatial separation of specific population subgroups within a wider population. Situations of segregation may take three forms: *enclaves*, in which congregation is characterized by internal harmony; *ghettoes*, which are more a product of discrimination than of congregation; and *colonies*, which are generally shorter lasting.

Traditional Patterns of Urban Structure

The organization of cities reflects the way they function. The typical U.S. city is based on several main elements of land use. The very center of the city, or **central business district (CBD)**, is the principal hub of shops and offices, together with some of the major institutional land uses such as the city hall, libraries, and museums. Normally the CBD has the densest concentration of shops and offices and contains the tallest buildings. The CBD is also a center of transportation connections, and thus usually contains the main rail and bus stations as well as the major hotels. (You can access many cities' downtown associations by going to the International Downtown Association website, linked through the textbook's website.) Surrounding the CBD is a **zone of transition**, which is an area of mixed commercial and residential uses, typically containing warehouses, specialized stores, small factories, apartment buildings, and older residential neighborhoods. Beyond this zone are residential neighborhoods and various kinds of suburbs.

Immigrant groups tend to move into neighborhoods with the lowest rent. Once they prosper, they move on, and the old neighborhood is taken over by another, later group of immigrants, resulting in waves of movement and distinct ethnic neighborhoods, such as the Chinatowns that are found in many cities. This process of ethnic neighborhood change is known as **invasion and succession**. Economic and social trends have changed the nature and appearance of cities and have led to various problems. Increased affluence in cities has led to a variety of lifestyles, and also to greater social polarization and ethnic division. Older residential neighborhoods are sometimes taken over by higher-income people, who renovate and restore many of the older buildings. This process is called **gentrification**. These processes have led to greater fragmentation, and to the relocation of different social groups to suburbs and edge cities.

The **fiscal squeeze** is a city problem resulting from increased limitations on city revenues, combined with increasing demands for expenditures. As wealthier households and industries relocate outside the central city boundaries, the city suffers a loss of tax revenue, while at the same time older neighborhoods may need increased investment in such things as transit lines, schools, and sewage systems. The decay of infrastructure—sewage systems, transit lines, water supplies—is a problem in many cities, as is poverty and the decay of urban neighborhoods. Landlords of older inner city apartments often have no incentives or money to repair or improve buildings, school systems decline because they are unable to attract good teachers and money for books and other supplies, and shops and services likewise suffer as neighborhoods decay.

Many poverty areas are also ethnic ghettoes. Though discrimination in housing is illegal in the United States, it may take place in subtle ways, such as through **redlining**, in which banks refuse to lend money to residents of "bad neighborhoods," resulting in a bias against minority neighborhoods. The idea of an **underclass** refers to a subset of the poor that is isolated from mainstream values and from the formal labor market, making it very difficult for these people to escape from poverty. A potential consequence of this extreme poverty is *homelessness*, a phenomenon that has been increasing in the world's large cities.

European cities often differ dramatically from American cities, and this is part of their appeal for American tourists. How and why do European cities differ? Many of the distinctive characteristics of European cities derive from their long histories. European cities are frequently characterized by: 1) *Complex street patterns*, which developed before the invention of the automobile, and in which streets are often narrow and set at unusual angles to each other; 2) *Plazas and squares*, which are typical of Greek, Roman, and Medieval cities; 3) *High density and compact form*, often because of a protective wall around the city or because of recent restrictions on suburban sprawl; 4) *Low skylines*, meaning that there are not many skyscrapers, partly because many central districts were built up before the invention of elevators, and partly because of building codes that required a cathedral or monument to be the highest structure in the city; 5) *Lively downtowns*, where the downtown is a center of social life and not just a place for offices; 6) *Neighborhood stability*, because Europeans do not move as often as Americans; 7) *The scars of war*, both because many European cities were built for defense and because many suffered damage in wars; 8) *Symbolism*, because of the presence of reminders of the past such as gothic cathedrals, palaces, and castles; and 9) *Municipal socialism*, which means that many residential buildings are owned by local governments. These features help account for the differences in appearance and feeling between European and American cities.

City planning and design have a long history. The Greeks and Romans, like the Chinese and Japanese, laid out many of their cities on a grid system. European urban planning emerged in the Renaissance, when rich and powerful regimes used urban design to symbolize their wealth, power, and destiny. Urban planners also tried to impose order, safety, and efficiency in the design of cities. Paris, for example, was completely redesigned between 1853 and 1870, during which many of the old, winding streets were replaced by a network of wide boulevards and public squares. Other cities followed the example of Paris, often making use of the **Beaux Arts** (pronounced "boh-zart") style, which combined Classical, Renaissance, and Baroque elements.

The **Modern movement**, a mid-20th-century movement in architecture and urban design, among other things, presented the idea that buildings and cities should be designed and run like machines. The movement's best known advocate was the Swiss architect Le Corbusier, whose designs attempted to dramatize technology, exploit industrial production techniques, and use modern materials and unornamented, functional designs. The International Style, which emphasized boxy glass and steel buildings, was a later part of the Modern movement.

Islamic cities are good examples of how social and cultural values influence urban form and the built environment. Islamic cities are normally characterized by a central mosque, a defensive wall around the city punctuated by city gates, by covered markets called *suqs*, and by houses with walled courtyards that protect the privacy of the inhabitants, especially the privacy of women. Take a virtual tour of an Islamic city by linking through the textbook's website to the IslamiCity site.

Problems of cities in the periphery include unemployment and underemployment. **Underemployment** is when people work less than full time even though they would prefer to work more hours. The lack of employment opportunities in many cities, especially those of the periphery, has led to the growth of the informal sector, in which people attempt to make a living as best they can, often by marginal activities such as selling souvenirs, shining shoes, writing letters for others, and scavenging. Often children are also involved in these occupations. The rapid growth of cities, especially of those in peripheral countries, has led to overcrowding and shortages of housing. Poor migrants to the city often live in shantytowns, squatter settlements, and slums. Some cities have responded to these settlement patterns through eviction, which usually leads to the construction of shantytowns in other parts of the city. Crime, poor health care, and poor sanitation often characterize these slums and shantytowns. You can examine a Venezuelan squatter settlement in detail at the Case Study of a Barrio in Caracas website, linked through the textbook's website.

Cities also face problems involving transportation. Increasing *automobility*—dependence on the automobile as the principal form of transportation—characterizes most cities in both core and peripheral countries. Traffic jams and accidents are commonplace in most large cities. Many cities have established or upgraded public transportation systems involving combinations of buses, trains, and light rail systems, but these are costly to maintain and construct, and it is difficult to wean people away from their cars. Other infrastructural problems, such as the provision of fresh water and the removal of wastes, are also characteristic of fast-growing cities.

Cities, and especially those in the periphery, face numerous environmental problems as well. Because of rapid population growth, inadequate sewage systems can lead to contamination of groundwater from human waste and from industrial effluents. Air pollution results from automobile exhausts and from industrial and power-generating activities. The burden of dealing with these problems rests on city governments. Problems result when city governments are highly fragmented, or when national governments allow little independent authority to municipal governments. Problems also result from financial constraints—the lack of adequate revenues combined with increased expenditures. Given these factors, the future of many cities appears grim.

New Patterns: The Polycentric Metropolis

Most cities have complex arrangements: for example, many cities contain *secondary business districts* in outer areas, and/or have *industrial districts* around factories and airports. Cities may also have edge cities on their borders. **Edge cities** are nodal concentrations of shopping and office space that are situated on the outer fringes of metropolitan areas, typically near major highway intersections. The Modern movement was often criticized for producing boring, austere, and dehumanizing buildings. A late 20th-century response to it, **postmodern urban design**, attempted to make buildings and city design more playful and interesting by incorporating new uses of color and ornamentation. This movement is still very much in vogue today.

The land use and organization of different parts of cities reflects the economic, political, and technological conditions prevailing at the time of a city's growth, and are partly a product of regional cultural values. Many of the most striking contrasts between cities of the world today are found between cities of the core and those of the periphery. These cities have different histories and face different problems today.

Review Questions

Some of the answers to these questions may be found only in the textbook, and not in the study guide.

1. Fast-growing suburban cities are sometimes known as _____.

2. A package of suburban land-use planning principles designed to curb sprawl is known as:
    a) redlining
    b) segregation
    c) smart growth
    d) defensible space

3. Protected enclaves of the rich, gentrified areas, ethnic enclaves, and excluded ghettoes are examples of _____.

City Spaces: Urban Structure

4. _____ was a style of urban design that sought to combine the best elements of all of the classic architectural styles.

5. The central nucleus of commercial land uses in a city is called the:
   a) zone in transition
   b) central business district (CBD)
   c) isotropic surface
   d) edge city

6. The territorial and residential clustering of specific groups or sub-groups of people is known as:
   a) invasion and succession
   b) defensible space
   c) congregation
   d) segregation

7. The _____ is the subset of the poor isolated from mainstream values and the formal labor market.

8. The area of mixed commercial and residential land uses surrounding a city's central business district (CBD) is called the _____.

9. The transmission of poverty and deprivation from one generation to another through a combination of domestic circumstances and local, neighborhood conditions is known as:
   a) underemployment
   b) invasion and succession
   c) gentrification
   d) cycle of poverty

10. True or False: Dualism refers to the juxtaposition in geographic space of the formal and informal sectors of the economy.

11. The single most dominant feature of the traditional Islamic city is the _____.

12. _____ are nodal concentrations of shopping and office space that are situated on the outer fringes of metropolitan areas, typically near major highway intersections.

13. _____ is a style characterized by a diversity of architectural styles and elements, often combined in the same building or project.

14. Increasing limitations on city revenues, combined with increasing demands for expenditure, is known as the _____.

True or False: "Boomburbs" refers to the invasion of older, centrally located working-class neighborhoods by higher-income households seeking the character and convenience of less expensive and well-located residences.

15. _____ is a process of neighborhood change whereby one social or ethnic group succeeds another.

16. Unlike most North American cities, many European cities are characterized by:
_____
_____
_____
_____

17. True or False: An isotropic surface is a hypothetical, uniform plain: flat, and with no variations in its physical attributes.

City Spaces: Urban Structure   181

18. Population subgroups that are seen—or that see themselves—as somehow different from the general population are called _____.

19. Metropolitan sprawl largely results from the development of this mode of transportation: _____.

20. The idea that buildings and cities should be designed and run like machines is called:
   a) Modern movement
   b) postmodern urban design
   c) Beaux Arts
   d) redlining

21. The practice whereby lending institutions delimit "bad risk" neighborhoods on a city map and then use the map as the basis for determining loans is called:
   a) defensible space
   b) smart growth
   c) redlining
   d) cycle of poverty

22. True or False: Gentrification is the spatial separation of specific population subgroups within a wider population.

23. When people work less than full time even though they would prefer to work more hours, it is called _____.

24. The original, core jurisdictions of metropolitan areas are called _____.

25. Gentrification refers to the invasion of older, centrally located working-class neighborhoods by _____-income households seeking the character and convenience of less-expensive and well-located residences.

© 2010 Pearson Education, Inc.

26. The three principal situations that can occur through congregation of specific groups of people in a city are:

_____
_____
_____

27. Street vending, shoe cleaning, craft work, scavenging, begging, and prostitution are all examples of the _____ economic sector.

Conceptual Questions

1. Describe the patterns of congregation in your community. Do different groups cluster in different areas? Why is this so?

_____
_____
_____
_____
_____
_____
_____
_____
_____

2. What activities characterize the central business district of your community? How is this district distinct from surrounding areas?

_____
_____
_____

3. Why do different activities cluster in different parts of cities? What models have been used to explain this clustering?

4. How do European cities differ from cities in North America?

5. What are some of the problems faced by cities in the periphery? How are these problems similar to and different from the problems of core cities?

6. What urban problems are most characteristic of your community? How did these problems arise? Can you offer any potential solutions to them?

12

Future Geographies

Chapter Objectives

The objectives of this chapter are to:

1. Examine both global and local prospects
2. Survey future resources, technology, and spatial change
3. Examine cultural dissonance and sustainability

Chapter Notes

Mapping Our Futures

What kind of future does the world hold? Predicting the future can be a risky and uncertain business, but that has not stopped people from trying. Scenarios of the future can be both optimistic and pessimistic. Optimists see the potential for technology to enable new ways of living and to solve many of the world's present problems. Pessimists note the finite nature of the Earth's resources, rapidly growing populations, and the failure of technology to solve many pressing problems. What is clear is that we have entered a period of transition and major change since 1989. Many unexpected events—such as the breakup of the Soviet Union, the unification of Germany, the end of apartheid in South Africa, and the bombing of New York's World Trade Center—have taken place in the few years since that date. (You can keep up to date on world events through the

Global Intelligence website, linked through the textbook's website.) The world has indeed become a very different place.

Resources, Technology, and Spatial Change

The future of the world system will depend on demands for natural resources and on the exploitation of new technologies. The availability of energy resources will be especially critical. By the year 2020, peripheral and semiperipheral countries are expected to account for more than half of world energy consumption, much of it driven by industrialization in those countries. Though the world does contain large reserves of energy resources, such as coal, oil, and natural gas, these reserves are not likely to be sufficient to meet growing demand. Energy prices may rise, shortages may appear, and the environment may suffer from increased exploration and exploitation. Conservation and technological advances may or may not affect the availability and nature of energy resources. It is difficult to predict the impact of new technologies on society and the economy. A few decades ago, no one really knew what the impact of such things as personal computers and the Internet would be. Today we know that they have revolutionized people's lives.

Technological advance is likely to continue, perhaps at an even greater pace. Transportation is one area where technology has been rapidly developing. High-speed rail systems, with trains that can reach speeds of 250 miles per hour, are now being developed and implemented. Intelligent Vehicle Highway Systems (IVHS) are part of the technology of smart cars, which use computers to assist their travel and travel routes along highways. And plans are in the works for much larger commercial aircraft as well as for faster airplanes—even some that may cut the travel time between Tokyo and Los Angeles down to two hours, meaning that people could conceivably commute from Asia to North America and vice-versa. Biotechnology is already a rapidly developing area of research. Genetic engineering has improved the quality of crops and domestic animals. Other applications will be industrial, including in the areas of waste disposal and biomass fuels.

Materials technologies are changing the way we use natural resources. New kinds of plastics, ceramics, and other synthetic materials may partially replace reliance on metals and other natural materials. Such technologies may have positive and negative effects, especially negative ones in countries that are major producers of metals and metal ores. Information technologies are already changing quickly (see the Silicon Valley to Internet Valley website, linked to the textbook's website, for more information). Most people in the core countries are aware of the rapid changes and improvements in computer technology, including software, as well as advances in communications through electronic mail and the Internet, and by satellite. Given the unpredictability of past advances, we probably have no idea of what the information technology future holds.

Regional Prospects

Globalization is having different effects on different kinds of world regions. We might speak about three general kinds of regions: the marginalized, the elite, and the embattled. The marginalized are those countries that are characterized by demographic, environmental, economic, and social stress. They are places that are becoming less important and less relevant to the global economy, and their future is uncertain. The elite, on the other hand, are the most prosperous core regions that can take advantage of globalization. The embattled are those places caught between the marginalized and the elite, and share some characteristics of both. The marginalized, elite, and embattled are not just regions or places, but are also individuals or groups that may be distributed in ways that do not coincide with national or regional boundaries.

Critical Issues and Threats

The immediate future will be characterized by a phase of geopolitical and geoeconomic transition; by the continued expansion of the world economy; and by the continued globalization of industry, finance, and culture. New technologies will be especially influential in shaping levels of economic development. The impact will not be entirely positive: changes of this kind will undoubtedly bring conflicts as well. The globalization of the economy is undermining the status of the territorial nation-state as the basic unit in the world system. Territorial nation-states will undoubtedly continue to exist, but their influence and importance may be weakened by transnational flows of information, capital, and culture. The world of the future may exist as a network of redundant state boundaries underneath a global economic, political, and cultural network. Residents of different countries may be less affected by national policies and more subject to international ones.

Cultural globalization is also having an impact on the contemporary world. A global consumer culture—largely originating in the United States—is evident throughout the core and peripheral countries. Certain products are especially symbolic of this process and these products can be found almost anywhere. Examples include Coca-Cola, Nike shoes, Levi's jeans, Marlboro cigarettes, and American television and films. This consumer culture can both destroy small local cultures and build bridges between cultures—American youth, for example, will have much in common with Russian youth, as both seek to acquire the same products, listen to the same music, watch the same films, and support the same sports teams.

Environmental problems are now being given greater consideration by core countries, and, to a lesser extent, by countries of the periphery. Many peripheral countries are exploiting their natural resources at a rapid rate, leading to various environmental problems. Many of these countries see it as their right to catch up with the development levels found in core countries. The globalization of environmental problems as well as their links with economic development programs has prompted much discussion about sustainable development (for more information about the social and cultural aspects of globalization, go to the Cultural Survival and Overcoming Human Poverty websites, linked through the textbook's website). Perhaps the greatest contribution of the concept of sustainable development is that it makes explicit links between environmental problems and development practices.

Much of the world's future will depend on trends in resource demand and technological innovation. Though it is difficult to predict the future, it appears that levels of economic development will continue to widen between core and peripheral countries, as well as within these countries. Within the core, countries such as the United States,

Japan, Germany and the European Union, China, and Russia will contend for a dominant role in the global economy. Understanding human geography will help one prepare for the coming global uncertainties.

Review Questions

Some of the answers to these questions may be found only in the textbook, and not in the study guide.

1. What has been identified as the main resource need for the future? _____.

2. What is meant by materials technologies?
_____
_____

3. _____ is the reason for the high proportion of energy consumption projected for peripheral and semiperipheral countries.

Conceptual Questions

1. What will the global landscape look like in 2020? What are some relative certainties, and some relative uncertainties?
_____
_____
_____
_____

2. How are changes in resource demand and technology likely to affect the world economy?

3. Do you think that the United States will continue to be the dominant force in the world economy? Why or why not? What country (or countries) is likely to take its place, and why?

4. What are some positive and negative impacts of the globalization of culture?

5. What is meant by sustainability? Can you give examples of sustainability initiatives in your own area?

Appendix 1

Review Questions Answer Key

Chapter 1: Geography Matters

1. d
2. cognitive images
3. accessibility
4. a
5. true
6. formal regions
7. geodemographic research
8. Global Positioning System (GPS)
9. social relations
10. b
11. infrastructure
12. region
13. globalization
14. regionalization
15. irredentism
16. true
17. sense of place
18. cognitive distance
19. spatial analysis
20. location, distance, space, accessibility, spatial interaction
21. c
22. identity
23. true

24. ordinary landscapes

25. regional geography

26. spatial interaction

27. regionalism

28. d

29. false

30. friction of distance

31. remote sensing

32. d

33. situation

34. geographical imagination

35. cognitive space

36. states

37. b

38. geographical information systems (GIS)

39. utility

40. localization economies

41. sense of place

42. latitude, longitude

43. spatial diffusion

Chapter 2: The Changing Global Context

1. colonialism

2. core, semiperiphery, periphery

3. false

4. a

5. masculinism

6. c

7. division of labor

8. true

9. environmental determinism

10. false

11. external arena

12. Middle East, South Asia, China, North and South America

13. fast world

14. c

15. hinterland

16. false

17. Great Britain, or United Kingdom, or England

18. import substitution

19. a

20. b

21. European Union, United States, Japan

22. neocolonialism

23. canals

24. c

25. transnational corporation

26. producer services

27. semiperipheral regions

28. slash-and-burn agriculture

29. false

30. spatial justice

31. c

32. true

33. c

34. true

35. world-system

36. United States, Japan

37. (answers vary)

Chapter 3: Geographies of Population

1. agricultural density
2. true
3. b
4. true
5. crude death rate (CDR)
6. false
7. demographic transition
8. d
9. dependency ratio
10. true
11. people, land area
12. a
13. true
14. push factors
15. forced migration
16. geodemographic analysis
17. Kenya, high birth rate
18. age–sex pyramid
19. c
20. guest workers
21. 62% (or about two-thirds)
22. b
23. infant mortality rate
24. false
25. life expectancy
26. medical geography
27. b
28. middle cohort

29. true

30. natural increase

31. international migration

32. food supply

33. a

34. nutritional density

35. census

36. c

37. true

38. suburbanization

39. false

40. total fertility rate (TFR)

41. voluntary migration

42. a

Chapter 4: Nature and Society

1. acid rain

2. siltation

3. Henry David Thoreau

4. Clovis point

5. greenhouse gas emissions

6. b

7. true

8. Rachel Carson

9. false

10. deforestation

11. harvesting of resources, emission of wastes in the manufacturing of goods and services, emission of wastes in the consumption of goods and services

12. d

13. industrialization, urbanization

14. c

15. b

16. ecosystem

17. c

18. global change

19. oil

20. true

21. conversion, modification

22. Paleolithic period

23. Impact, Population, Affluence, Technology

24. preservation

25. demographic collapse

26. a

27. society

28. environmental ethics

29. false

30. Transcendentalism

31. b

32. political ecology

33. Columbian Exchange

Chapter 5: Cultural Geographies

1. cultural complex

2. Asia

3. cultural geography

4. c

5. cultural landscape

6. cultural nationalism
7. Islam
8. false
9. race
10. culture
11. a
12. tribe
13. b
14. ethnicity
15. Indo-European
16. false
17. *genre de vie*
18. c
19. Islamism
20. religion, language
21. H.C. Darby
22. language
23. language branch
24. false
25. language group
26. Muslim
27. religion
28. true
29. c
30. false
31. diaspora
32. cultural system
33. historical geography
34. sexuality
35. gender

Chapter 6: Interpreting Places and Landscapes

1. cosmopolitanism
2. range of perspectives in social inquiry, artistic expression, political empowerment
3. c
4. Modernism (or Modernity)
5. humanistic approach
6. consumption
7. false
8. cross
9. a
10. true
11. b
12. sacred space
13. landmarks
14. a
15. territoriality
16. symbolic
17. true
18. individual
19. written, read
20. social interaction, access to people and resources, focus and symbol of group membership and identity
21. ethology
22. semiotics

Chapter 7: The Geography of Economic Development

1. a
2. ancillary activities
3. wider
4. true
5. sustainable development
6. c
7. carrying capacity
8. false
9. d
10. ecological footprint
11. false
12. elasticity of demand
13. Henry Ford
14. conglomerate
15. gross domestic product (GDP), gross national product (GNP), purchasing power parity (PPP)
16. true
17. a
18. secondary activities
19. external economies
20. world cities
21. foreign direct investment
22. tourism
23. gross domestic product (GDP), gross national product (GNP)
24. true
25. b
26. d
27. agglomeration diseconomies
28. (answers vary)
29. Organization for Economic Cooperation and Development
30. Reaganomics

31. terms of trade

32. Western Europe, North America, Former Soviet bloc, Japan and East Asia

33. infrastructure (or fixed social capital)

34. conglomerate corporations

35. initial advantage

36. true

37. (answers vary)

38. quaternary

39. a

40. c

41. transnational corporations

42. geographical path dependence

43. *maquiladoras*

Chapter 8: Agriculture and Food Production

1. aquaculture

2. ejidos

3. famine

4. food security

5. true

6. land reform

7. agrarian

8. mechanization, chemical farming, food manufacturing

9. agribusiness

10. agricultural industrialization

11. biotechnology

12. false

13. tropical rainforests

14. true

15. crop rotation
16. double cropping
17. food chain
18. a
19. Asia
20. food regime
21. globalized agriculture
22. increase global agricultural productivity (or food supply)
23. Green Revolution
24. b
25. intensive subsistence agriculture
26. international grain markets (or prices)
27. mechanization
28. b
29. shifting cultivation
30. true
31. c
32. swidden
33. d
34. subsistence
35. biotechnology

Chapter 9: The Politics of Territory and Space

1. intifada
2. United States
3. (answers vary)
4. regionalism
5. Zionism
6. redistricting

7. globalization, interconnectedness of place
8. false
9. a
10. decolonization
11. frontier
12. d
13. Australia, New Zealand
14. East/West divide
15. colonialism, imperialism
16. regionalism (or nationalism, or terrorism)
17. c
18. c
19. international organization
20. false
21. nation-state
22. false
23. geopolitics
24. nationalism
25. true
26. reapportionment
27. a
28. sovereignty
29. territorial organization
30. false
31. c
32. centrifugal
33. federal
34. true
35. democratic rule
36. youth, maturity, old age
37. nation

Chapter 10: Urbanization

1. Middle East
2. $1/n$
3. reurbanization
4. b
5. rank-size rule
6. urban form
7. central place theory
8. Tokyo
9. core
10. true
11. colonial city
12. industrialization
13. false
14. b
15. true
16. megacity
17. primate
18. overurbanization
19. North America
20. b
21. (answers vary)
22. gateway
23. squatter settlements
24. c
25. true
26. d
27. central place

Chapter 11: City Spaces: Urban Structure

1. boomburbs
2. c
3. socio-spatial formations
4. Beaux Arts
5. b
6. c
7. underclass
8. zone in transition
9. d
10. true
11. principal mosque, or *Jami*
12. edge cities
13. postmodern urban design
14. fiscal squeeze
15. false
16. (answers vary)
17. true
18. minority groups
19. automobile
20. a
21. c
22. false
23. underemployment
24. central cities
25. higher
26. enclaves, ghettoes, colonies
27. informal

Chapter 12: Future Geographies

1. energy

2. (answers vary)

3. industrialization

Appendix 2

Writing a Term Paper

Writing a good paper is not really all that difficult if you prepare and organize well. The first essential issue is deciding on a **topic**. Your topic may be assigned, or you may have some freedom of choice in selecting it. A good way to approach a term paper topic is to think about the point you are trying to make. A good paper does not just describe an issue in the way that an encyclopedia does. A good paper will have a **point or argument**, and it will show something about what *you* think about the issue. A good paper will combine thorough research, which provides supporting evidence, and your own ideas, which provide the argument or thesis of the paper. You might approach the point you are trying to make by thinking of it as a question that your paper will help answer.

The second essential issue in writing a good paper is good organization—both before and during the writing of the paper. Writing a long paper will seem much easier—and like much less work—if you prepare a good **outline** first. Your outline should contain all of the main points you want to make. Write out these points first, then arrange them in a way that makes sense and seems to flow. If you can write a few paragraphs for each point, you will find that your paper almost writes itself.

Papers should follow the traditional arrangement of introduction, body, and conclusion. The **introduction** will announce your topic and the point you are trying to make. The **body** of the paper will present your data and evidence for that point, and elaborate the argument; it may be divided into separate sections. The **conclusion** will sum up the paper and once again remind the reader of the point you have made.

A final consideration is the use of **references**. Where have you found all of the information you've used in writing your paper? There are many ways to present references, and below you'll find two of the most common ways explained. References (or citations) are all of those sources that are mentioned somewhere in your paper. A

bibliography, on the other hand, is a list of all sources consulted, even though you may not mention them in the text of your paper.

If you have any doubts or questions about how to write a good paper, ask your instructor.

Using References in Term Papers

References (also known as **citations**) indicate the sources of material you are using for your paper. References may be books; journal, magazine, or newspaper articles; government documents; personal interviews; and websites. For **books** you need to indicate the author and title of the book, the city of publication (also indicate the state or country if it's not obvious), the name of the publisher, the year of publication, and who translated the book (if it's a translation). Using encyclopedias for anything other than basic facts and statistics is generally frowned upon. For **chapters in books** you need to indicate the author of the chapter, the chapter title, the chapter's page numbers, the editors of the book, the book's title, and the publisher and date and place of publication. For **journal, magazine, and newspaper articles** you need to indicate the title of the article and its author (some newspaper articles don't list an author—for those you can just use the title), the name of the journal, magazine, or newspaper, and the volume number, date of publication, and page numbers. For **government documents** you should follow the example of an article, but give any reference number for the document as well as the name of the agency that published it. For **personal interviews**, you should list the name of the person you interviewed, their job or title, and the date and place (city and state) of the interview. If you interview somebody by e-mail or telephone, you can mention this instead of the place. For **websites**, you should list the title of the web page, the URL (web address), and the date you accessed it.

You need to reference material if you are making an **exact quote** from someone else's work, and also if you are using their **idea or concept** (even if it's paraphrased and not in their exact words). If you are listing any **facts or figures** that are not commonly known, then you also need to reference where you got that information. If you use somebody else's map or illustration, you also need to indicate the source for that.

References are normally listed in one of two general ways. The first is by using notes, which are called **footnotes** when they appear at the bottom of the page, and are called **end notes** when they are all listed at the end of the paper. Notes are indicated by a small superscript number in the body of the paper. The information about the source is then listed either at the bottom of the page or at the end of the paper. The second way of listing references is known as the **author-date system** (sometimes known as the Harvard system). If you use this system, instead of using a superscript number, you indicate the author's last name, year of publication, and page number (if appropriate) in parentheses just after the material quoted or cited. You then list all the references at the end of the paper, in alphabetical order. See the examples below.

There are many variations of these two systems. In general, you can use any accepted style, such as one that appears in a major journal or university press publication. Whatever style you choose, be consistent throughout the paper. Do not mix styles.

Maps and illustrations are often useful in making a point in a paper, as are things like **tables and graphs** if you have a lot of statistical data. The use of these will depend on the topic of your paper. Be sure to give the source of any maps, tables, or graphs that you use.

Using footnotes (bottom of page) or end notes (all notes at end of paper):

Examples of how the citations would appear in a paper or essay:

1. The reality of America "is created and maintained by overarching myths."[1]

2. The relationship between indigenous peoples and government in northern Canada is different from their relationship in other parts of the country.[2]

3. R. Gerard Ward argues that Europeans and Pacific Islanders have had different perceptions of the Pacific region.[3]

4. Polar regions play a key role in the operation of the global atmospheric system.[4]

5. The Australian National University was founded by the Australian Government in 1946.[5]

6. The classification of pandas as either members of the bear or raccoon family is still disputed.[6]

Examples of how footnotes would appear (at the bottom of the page) or how end notes would appear (in a section called "Notes" at the end of the paper) (notes are <u>listed in the order in which they appear</u>):

[1] Yi-Fu Tuan, *Cosmos and Hearth: A Cosmopolite's Viewpoint* (Minneapolis: University of Minnesota Press, 1996), p. 5.

[2] Sanjay Chaturvedi, *The Polar Regions: A Political Geography*, (Chichester, England: John Wiley and Sons, 1996), p. 149.

[3] R. Gerard Ward, "Earth's Empty Quarter? The Pacific Islands in a Pacific Century" *The Geographical Journal* 155, no. 2 (1989): 235–246.

[4] Chaturvedi, p. 21.

[5] Australian National University, About ANU, http://info.anu.edu.au/Discover_ANU/About_ANU/Profile/index.asp, February 27, 2005.

[6] Michael Pretes, "Panda." In *Encyclopedia of Modern Asia*, edited by David Levinson and Karen Christensen, Vol. 4, p. 459 (New York: Scribner, 2002).

Using the author-date system (the Harvard system):

Examples of how the citations would appear in the paper or essay:

1. The reality of America "is created and maintained by overarching myths" (Tuan 1996: 5).

2. The relationship between indigenous peoples and government in northern Canada is different from their relationship in other parts of the country (Chaturvedi 1996: 149).

3. R. Gerard Ward (1989) argues that Europeans and Pacific Islanders have had different perceptions of the Pacific region.

4. Polar regions play a key role in the operation of the global atmospheric system (Chaturvedi 1996: 21).

5. The Australian National University was founded by the Australian Government in 1946 (Australian National University 2005).

6. The classification of pandas as either members of the bear or raccoon family is still disputed (Pretes 2002).

Examples of how the citations would appear in the references, in a separate section of the paper called "References" (they are listed <u>in alphabetical order</u>):

© 2010 Pearson Education, Inc.

Australian National University. 2005. About ANU.
 http://info.anu.edu.au/Discover_ANU/About_ANU/Profile/index.asp, February 27, 2005.

Chaturvedi, Sanjay. 1996. *The Polar Regions: A Political Geography*. Chichester, England: John Wiley and Sons.

Pretes, Michael. 2002. "Panda." In *Encyclopedia of Modern Asia*, edited by David Levinson and Karen Christensen, Vol. 4, p. 459. New York: Scribner.

Tuan, Yi-Fu. 1996. *Cosmos and Hearth: A Cosmopolite's Viewpoint*. Minneapolis: University of Minnesota Press.

Ward, R. Gerard. 1989. "Earth's Empty Quarter? The Pacific Islands in a Pacific Century." *The Geographical Journal* 155 (2): 235–246.